Technische Universität Dresden
Fakultät Elektrotechnik und Informationstechnik

GATE STACK ENGINEERING FOR EMERGING POLARIZATION BASED NON-VOLATILE MEMORIES

Milan Dragan Pešić

Born in 14.04.1988, SFR Yugoslavia

A dissertation for partial fulfillment of the requirements for the degree of Doktoringenieur (**Dr.-Ing.**)

(genehmigte Dissertation)

Chairman:	Prof. Dr. J.W. Bartha (TU Dresden)		
Supervisor:	Prof. Dr. T. Mikolajick (TU Dresden)		
Supervisor:	Prof. Dr. E. P. Burte (OVGU Magdeburg)	Date of submission:	02.02.2017
4th member:	Prof. Dr. habil. J. Lienig (TU Dresden)	Date of defense:	24.05.2017

Abstract

The hafnium based ferroelectric memories offer a low power consumption, ultra-fast operation, non-volatile retention as well as the small relative cell size as the main requirements for future memories. These remarkable properties of ferroelectric memories make them promising candidates for non-volatile memories that would bridge the speed gap between fast logic and slow off-chip, long term storage. Even though the retention of hafnia based ferroelectric memories can be extrapolated to a ten-year specification target, they suffer from a rather limited endurance. Within the lifetime of the device, the two main stages, wake-up and fatigue, can be identified. Up to now, the mechanisms responsible for these two performance-limiting stages have not been revealed. Therefore, this work targets relating the field cycling behavior of hafnia based ferroelectric memories to the physical mechanisms taking place within the film stack. Establishing a correlation between the performance of the device and underlying physical mechanisms is the first step toward understanding the device and engineering guidelines for novel, superior devices.

In the frame of this work, first, the complexity of the interplay between charge trapping and ferroelectric switching was decoupled by using a non-switching ZrO_2 based dynamic random access memories (DRAM) as a test vehicle. The ZrO_2 based devices were chosen due to the fact that HfO_2 and ZrO_2 are considered to be twin oxides with similar band-structure properties. After a detailed electrical characterization and charge transport study without ferroelectricity-induced parasitic, the study was extended to ferroelectric HfO_2.

In the second part, the focus was on identification of the root cause for the increasing remnant polarization during the wake-up phase and its subsequent degradation with further cycling. An in-depth ferroelectric and dielectric characterization, analysis and TEM study was combined with comprehensive modeling approach. Drift and diffusion based vacancy redistribution was found as the main cause for the phase transformation and consequent increase of the remnant polarization. From the experimentally obtained defect evolution with ferroelectric switching and device modeling, a detailed understanding of the main mechanisms behind the evolution of the ferroelectric response was derived.

Finally, based on Landau theory, a simple way to utilize the high endurance strength of anti-ferroelectric (AFE) materials and achieve non-volatility in state-of-the-art ZrO_2 based DRAM stacks was proposed. By employing electrodes with different workfunctions, an internal bias field is introduced within the anti-ferroelectric stack, thus creating two stable non-volatile states. Moreover, the fabrication of the world's first non-volatile AFE-RAM is reported. These findings represent an important milestone and pave the way toward a commercialization of (anti)ferroelectric non-volatile memories based on simple binary-oxides.

Kurzzusammenfassung

Ferroelektrische Speicher auf Basis von Hafniumoxid bieten nicht nur einen niedrigen Energieverbrauch, ultraschnellen Zugriff, nichtflüchtige Datenhaltung sondern auch eine geringe relative Zellgröße als die Hauptanforderungen an zukünftige Speicher. Obwohl die Datenhaltung hafniumoxidbasierter ferroelektrischer Speicher bis zu einem Spezifikationsziel von zehn Jahren extrapoliert werden kann, leiden sie unter einer begrenzten Zyklenfestigkeit. Innerhalb der Lebensdauer der Speicher lassen sich zwei Regime unterschieden, die „Wake-up" und „Fatigue" genannt werden. Die vorliegende Arbeit zielt darauf ab, genau diese Instabilitäten während des zyklischen Betriebes der Speicherelemente hinsichtlich ihrer physikalischen Ursachen innerhalb des Schichtstapels zu untersuchen. Eine Korrelation zwischen dem Bauelement und zugrundeliegender physikalischer Effekte herzustellen, ist der erste Schritt hin zu einem Verständnis des Bauelementes und zur Ableitungen von Optimierungsansätzen.

Im Rahmen dieser Arbeit musste zunächst die Komplexität des Zusammenspiels zwischen Elektronen-Einfang und ferroelektrischem Schalten aufgelöst werden. Dies gelang durch die Verwendung eines nichtschaltenden Dynamic Random Access Memories (DRAM) auf der Basis von ZrO_2 als Testvehikel. Die ZrO_2-basierten Bauelemente wurden aufgrund der Tatsache gewählt, dass HfO_2 und ZrO_2 als Zwillingsoxide betrachtet werden, die sehr ähnliche Bandstruktureigenschaften aufweisen. Nach einer detaillierten elektrischen Charakterisierung und Studie zum Ladungstransport ohne ferroelektrizitätsinduzierte parasitäre Effekte wurde die Studie auf ferroelektrisches HfO_2 ausgeweitet.

Im zweiten Teil lag der Fokus auf der Identifizierung der Ursache für die zunächst ansteigende remanente Polarisation („Wake-up") sowie ihre anschließende Abnahme („Fatigue") mit weiteren Schaltzyklen. Eine eingehende ferro- und dielektrische Charakterisierung sowie eine TEM-Studie wurden mit umfassender Modellierung kombiniert. Als Hauptursache für die Phasentransformation und die daraus resultierende Erhöhung der remanenten Polarisation wurde eine drift- und diffusionsbasierte Vakanzen-Umverteilung identifiziert. Aus der experimentell bestimmten Defektevolution und der Modellierung des ferroelektrischen Schaltens sowie des gesamten Bauelementes wurde ein detailliertes Verständnis der Mechanismen abgeleitet, die sich hinter der Evolution des makroskopischen ferroelektrischen Verhaltens verbergen.

Schließlich wurde auf der Grundlage der Landau-Theorie ein einfacher Weg zur Nutzung der hohen Zyklenfestigkeit von anti-ferroelektrischen (AFE) Materialien und zur Erzielung nicht-flüchtiger Speicherung in modernen ZrO_2-basierten DRAM-Stapeln vorgeschlagen. Durch die Verwendung von Elektroden mit unterschiedlicher Austrittsarbeit wird ein internes Feld innerhalb des anti-ferroelektrischen Stapels induziert, wodurch zwei stabile, nicht-flüchtige Zustände erzeugt werden. Darüber hinaus wird die Herstellung des weltweit ersten nicht-flüchtigen AFE-RAMs demonstriert. Diese Ergebnisse stellen einen wichtigen Meilenstein dar und ebnen den Weg zu einer Kommerzialisierung von (anti-) ferroelektrischen nicht-flüchtigen Speichern auf Basis einfacher binärer Oxide.

Acknowledgments

At the very beginning I would like to express my gratitude to the whole NaMLab "gang", who very quickly and kindly accepted me in the team. The working atmosphere of the whole crew made me go to work with the smile on my face. Thank you, guys-you were great, but I have to admit that topic was fantastic as well.

Among the whole crew, very special thanks belong to Uwe Schroeder and Stefan Slesazeck who were my daily supervisors and always kept me on track. Besides keeping the eye on the big picture the freedom I received from you guys made this work possible.

I am very thankful to my doctoral advisor, Thomas Mikolajick for giving me opportunity to join NaMLab and realize many ideas which I had. Further, I am very grateful to Prof. Mikolajick for supervising this thesis from the side of Dresden University of Technology and for being a pleasant dialog partner in all scientific and non-scientific matters and most important giving me the opportunity to learn many new practical skills that will benefit my further carrier.

Prof. Luca Larcher is gratefully acknowledged for enabling me opportunity to use MDLab software, implement my ides within it and work with his fantastic team. Everett D. Grimley, Xiahan Sang and James M. LeBeau (NC State University, Raleigh) as well as KU Leuven collaboration partners Nadiia Kolomiets and Valeri Afanas'ev are gratefully acknowledged for their great contribution to this work by performing the TEM studies and IPE measurements, respectively. Samsung, Qimonda and GlobalFoundries are acknowledged for providing the test hardware which I characterized extensively.

My special gratitude goes to the people with whom I fabricated, measured, discussed, wrote papers, proofread the thesis, complemented with ideas etc... Claudia, Tony, Steve, Micha, Halid, Johannes, Franz, Ekaterina thank you for all. Exceptional thanks and respect goes to Claudi who was always believing in every crazy idea I simulated and provided me with as many samples as I was able to destroy by heavy characterization. Very special thanks go to Everett and Tony who enriched almost article-free beast with an American English and German English flavor, respectively.

My deep gratitude goes to my friends and girlfriend for being full of understanding for all the time spent with TCAD and MATLAB and for all: "I have to measure this weekend"; "Just a minute I want to simulate one effect"; "Wait, wait I got an interesting idea, I have to check it now" sentences which they heard quite often.

Finally, my parents Vera and Dragan are gratefully acknowledged from being a lifelong support.

Contents

Abstract I
Kurzzusammenfassung II
Acknowledgments III
List of Symbols VII
Abbreviations X
1   Introduction 1
2   Fundamentals 5
2.1   High-k Materials 5
2.2   Charge Transport and Leakage Mechanisms 8
2.2.1 Defectless transport mechanisms 9
2.2.2 Defect-mediated transport mechanisms 10
2.3   Phase Transitions and Ferroelectricity within the High-k Materials 14
2.3.1 Theory of the (Anti-)ferroelectric Physics 15
2.3.2 Models for Accounting Ferroelectric Properties 16
2.4   Ferroelectricity in HfO_2 19
2.5   Ferroelectric Memories 22
2.5.1 Ferroelectric Memory History and Architecture Overview 22
2.5.2 Hafnium-Based Ferroelectric Memories 25
3   Description of the Devices Under Test 27
3.1   Metal-Insulator-Metal (Semiconductor) Capacitor Structure 27
3.1.1 ZrO_2 Based Capacitors 27
3.1.2 Doped HfO_2 Based Capacitors 29
3.2   Ferroelectric Field Effect Transistor (FeFET) 29
4   Characterization Methods 30
4.1   Dielectric Characterization Methods 30
4.1.1 Leakage Current Defect Spectroscopy 30
4.1.2 Dielectric Absorption Test (DA) 31
4.1.3 Breakdown Voltage Test (VBD) 32
4.2   Ferroelectric Characterization Methods 33
4.2.1 Polarization-Voltage Test 33
4.2.2 PUND Test 33
4.2.3 First Order Reversal Curve Test 34
4.3   Memory Specific Characterization Methods 36
4.3.1 Endurance Test 36
4.3.2 Retention Test 36
5   Capacitor Stack Properties and their Influence on the Charge Transport 38
5.1   Basic Properties of ZrO_2 based Metal-Insulator-Metal Capacitors 39
5.2   Modeling of the Charge Transport within ZrO_2 and HfO_2 Based High-k Dielectrics 43
5.2.1 Determining Parameters Governing Charge Transport within ZrO_2 and HfO_2 High-k Dielectrics 43
5.2.2 Charge Transport Modeling Using a Compact Modeling Approach 48
5.2.3 Charge Transport Modeling Using a TCAD Modeling Approach 53

5.2.4 Charge Transport Model of Sr:HfO$_2$ MIM Capacitor Developed by MDLab Software 57
5.3   Introduction of Internal Bias Fields 59
5.4   Summary 60
6   Field Cycling of the Ferroelectric High-k Materials 62
6.1   Wake-up Behavior in the Ferroelectric Memories 63
6.1.1 FORC and Internal Bias Screening 65
6.1.2 Structural Changes and Transmission Electron Microscopy Study 67
6.2   Modeling of the Wake-up Behavior 73
6.3   Polarization Fatigue and Dielectric Degradation 80
6.4   Road Towards 1T Ferroelectric Memory-FeFET 84
6.4.1 Electric Field Cycling Behavior of FeFET 85
6.4.2 Modeling of the FeMIS and FeFET Stack 89
6.5   Summary and Outlook 93
7   Anti-ferroelectric Non-volatile Memory 96
7.1   On how Classic DRAM Material ended up Anti-ferroelectric 97
7.2   Stabilization of the Tetragonal Phase in ZrO$_2$ Thin Films 98
7.3   Intrinsic Properties of Anti-ferroelectric Materials 101
7.4   Building a Anti-ferroelectric Non-volatile Memory 103
7.4.1 Theoretical Analysis and Modeling of AFE-RAM 103
7.4.2 Practical Realization of the AFE-RAM 106
7.5   Reliability and Device Performance 109
7.5.1 Retention and Endurance of AFE-RAM 109
7.5.2 Field Cycling Behavior and Properties of the AFE-RAM 110
7.5.3 Exploring the Limits of ZrO$_2$-based AFE-Memories 112
7.5.4 Temperature Stability of the AFE-based Non-volatile Memories 114
7.6   Summary 116
8   Summary and Outlook 117
Bibliography 120
Resume 132

List of Symbols

Symbol	Description
k	Dielectric constant
A	Area
A^{**}	Richardson's constant
B	Bond polarization factor
c	Proportionality constant
C	Capacitance
C_{BIL}	Capacitance of the bottom interface
CBO	Conduction band offset
C_{eq}	Equivalent capacitance
C_{FE}	Capacitance of ferroelectric
C_{IF}	Interface capacitance
C_{IFsem}	Capacitance of the interface and semiconductor combined
c_i^n	Capture rate
C_{max}	Maximal value of capacitance
C_{meas}	Measured Capacitance
C_{min}	Minimal value of capacitance
C_{TIL}	Capacitance of top interface
D	Dielectric displacement vector
d	thickness
d_{FE}	Thickness of the ferroelectric
d_{IF}	Thickness of the interface
E	Electric field
$E_{A,D}$	Activation energy for diffusion
$E_{A,G}$	Activation energy for defect generation
$E_{A,R}$	Activation energy for recombination
$E_{BandGap}$	Band gap energy
E_{BD}	Breakdown field
E_{bias}	Bias field
E_{built_in}	Internal bias field
E_c	Coercive field
E_{dep}	Depolarization field
$E_{external}$	External electric field
e_i^n	Emission rate
E_r	Reversal electric field
E_s/E_{BS}	Electric field required for switching and backswitching of polarization
E_{static}	Static electric field
$f^{n/p}$	Trap occupation function
G	Generation rate
h	Planck's constant
$\hbar$	Reduced Planck's constant
I	Current
i	Number of capture of emission process
I_{sense}	Sensing (monitor) current
J	Current density
J_{DT}	Current density due to the direct tunneling
J_{FN}	Fowler-Nordheim current density
J_{SE}	Schottky emission current density
J_{SE}	Current density due to the Schottky emission
J_{TAPF}	Current density due to the trap assisted Poole-Frenkel
J_{TAT}	Current density of trap assisted current component

k_B	Boltzmann's constant
k_{BIL}	Capacitance of the bottom interface
k_{FE}	Capacitance of ferroelectric
k_{nonlin}	Material dependent proportionality constant
k_{ox}	Dielectric constant of the oxide
k_{SiO2}	Dielectric constant of SiO_2
k_{TIL}	Capacitance of top interface
m^*	Effective mass of the electron
m_e	Mass of the electron
N_t	Trap density
P	Dielectric polarization
P_{eff}	Effective polarization
P_{In}	Injection probability
P_{nonlin}	Polarization denoting material nonlinearity
P_{off}	offset polarization
P_{pol}	Transient polarization term of Preisach model
P_r	Remnant polarization
P_{r_rel}	Relaxed polarization
P_s	Saturation polarization
P_{sp}	Spontaneous polarization
P_{TAT}	trap-to-trap tunneling probability
q	Elementary charge
Q	Generalized coordinate
Q_f	Charge
$Q_{trapped}$	Trapped charge
R_D	Diffusion rate
r_i^n	Capture and emission function
R_R	Recombination rate
s	Distance between electron traps
T	Temperature
T_c	Phase transition temperature
U_b	Defect potential energy (Occupied energy oscillator)
U_f	Defect potential energy (Empty energy oscillator)
V	Applied voltage
V_{FB}	Flatband voltage
V_o	Neutral oxygen vacancy
V_o^+	Positive oxygen vacancy
V_o^{++}	Two-fold positive oxygen vacancy
V_{th}	Threshold voltage
WF	Workfunction
W_{opt}	Optical ionization energy of the defect
W_t	Thermal ionization energy of the defect (Trap depth)
α, β, γ	Landau expansion coefficients
ΔV	Voltage shift
ε_0	Vacuum permitivity
ε_{opt}	Optical permittivity
ε_r	Relative dielectric permittivity
Θ	Curie temperature
λ	Jump distance
v	Bond vibration frequency
ρ	Preisach switching density
τ_{eff}/τ_p	Material specific time constants
Φ	Free energy
Φ_{AFE}	Free energy of the antiferroelectric
Φ_{ms}	Workfunciton difference between semiconductor and metal
χ	Dielectric susceptibility

Abbreviations

Abbreviation	Description
1T	One transistor
1T-1C	One transistor, one capacitor
2D	Two dimensional
3D	Three dimensional
AC	Alternating current
AFE	Anti-ferroelectric
AFE-RAM	Anti-ferroelectric random access memory
ALD	Atomic layer deposition
BD	Breakdown
BE	Bottom electrode
BL	Bit line
BTO	Barium titanate
CB	Conduction band
CBO	Conduction band offset
CET	Capacitance equivalent thickness
CMOS	Complementary metal oxide semiconductor
CPU	Central processing unit
C-V	Capacitance-voltage
CVD	Chemical vapor deposition
CVS	Constant voltage stress
DA	Dielectric absorption (relaxation)
DC	Direct current
DHM	Dynamic hysteresis measurement
DRAM	Dynamic random access memory
DRO	Destructive read-out
DT	Direct tunneling
DUT	Device under test
EOT	Equivalent oxide thickness
ERS	Erase
EWF	Effective workfunction
FE	Ferroelectric
FeCap	Ferroelectric capacitor
FeFET	Ferroelectric field effect transistor
FeMOS	Ferroelectric metal oxide semiconductor
FeRAM	Ferroelectric random access memory
FFE	Field-induced ferroelectric
FN	Fowler-Nordheim
FORC	First order reversal curve
GB	Grain boundaries
HDD	Hard disk drive
HF	Hydrofluoric acid
HKMG	High-k metal gate
IBL	Interface buffer layer
IF	Interface
IoT	Internet of Things
IPE	Internal photoemission spectroscopy
I-V	Current-voltage
kMC	Kinetic Monte Carlo
LCDS	Leakage current defect spectroscopy
LGD	Landau-Ginzburg-Devonshire
M	Monoclinic
MFIS	Metal ferroelectric isolator metal
MFIS-FET	Metal-ferroelectric-insulator-semiconductor field effect transistor
MFM	Metal-ferroelectric-metal
MFS	Metal-ferroelectric-semiconductor
MIM	Metal-isolator-metal
MLC	Multi-level cell
MRAM	Magnetic random access memories
MW	Memory window
NDRO	Non-destructive read-out
NSS	New same state
NVM	Non-volatile memory
O	Orthorhombic
OS	Opposite state
O-vacancy	Oxygen vacancy
PCM	Phase change memory
PE	Paraelectric
PF	Poole-Frenkel emission
PL	Plate line
PRG	Program
PTO	Lead titanate
PUND	Positive up negative down
P-V	Polarization-voltage
PVD	Physical vapor deposition
PZT	Lead zirzonate titanate
RAM	Random access memory
RRAM	Resistive random access memories
SBH	Schottky barrier height
SC-1	Standard clean 1
SE	Schottky emission
SEM	Scanning electron microscopy
SILC	Stress induced leakage current
SRAM	Static random access memory
SS	Same state
STEM	Scanning transmission electron microscopy
STT-MRAM	Spin transfer torque magnetic random access memories
T	Tetragonal
TAPF	Trap-assisted Poole Frenkel
TAT	Trap-assisted-tunneling
TCAD	Technology Computer Aided Design
TE	Top electrode
TEM	Transmission electron microscopy
TEMAZr	Zr based metal organic precursor
TMA	Trimethylaluminum
UHV	Ultra-high vacuum
UniMoRE	University of Modena and Regio Emillia
VB	Valence band
VBD	Breakdown voltage test
VM	Volatile memory
WF	Workfunction
WKB	Wentzel, Kramers and Brillouin
WL	Word line
XRD	X-ray diffraction
ZAZ	$ZrO_2/Al_2O_3/ZrO_2$

1 Introduction

On the course of history of mankind different methods have been used for preservation of the information. One of the first methods for storing the information is the writing system. *Per definition* the writing system is a method of visual representation of the verbal communication. At the same time, it represents a system used for the information storage i.e. a memory. One of the earliest writing systems and storage memories is the cuneiform (wedge shaped) script. Here, information is stored by wedge-shaped marks on clay tablets. Further, a group of symbols represents a word which, in computer science is being defined as the natural unit of data operated by certain processor design and its instruction set. Apart from other system components, any system containing information processing unit consists of the elements that are used for temporary and permanent information storage or memory.

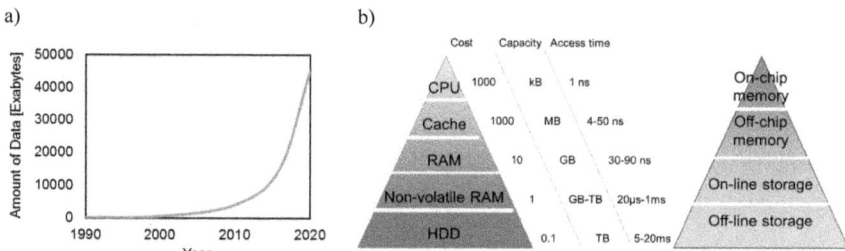

Figure 1.1 Memory requirements. a) Amount of data created or copied annually. The data are from ref. [1,2]. b) Memory hierarchy pyramid comparing the cost per bit, access time and storage density.

Memories can be differentiated by various criteria. E.g. according to the ability to retain a certain logic state after the removal of the external excitation, one can distinguish between non-volatile (NVM) and volatile memories (VM). In today's modern society or silicon era, complementary metal oxide semiconductor (CMOS) [3] devices that can be both NVMs and VMs are used for storing the binary state of logical "1" and "0". Besides the binary state memories, which discriminate between only two states, multilevel cells (MLC) provide ability to store more bits within the same cell enabling higher storage densities. The significance of the storage density can be seen in Figure 1.1a. The Exabyte (EB) as a unit is not used in the everyday language, but it is an important unit to be defined since each day around 5 - 8 Exabyte of new data are generated [1,2]. 1 EB represents 1 million of terabyte or 10^{18} Byte (1 Byte = 8 Bits). The most alarming fact is that 90 % of global data has been generated within the last two years. Data servers became so huge that they account for a significant portion (around 7 %) of the world's total electricity consumption [1]. This trend together with the recent expansion of the internet-of-things (IoT) market makes the situation even more complex.

1 INTRODUCTION

Besides the storage density and power consumption, the third most important requirement is the access time. In the Figure 1.1b memory hierarchy is given. A huge gap between the logic and central processing unit (CPU) with respect to the working dynamic random access memory (DRAM) as well as long term storage NVMs like hard disk drives (HDD) can be seen. Therefore, the performance of the whole computer system heavily depends on its memory performance. This gains even bigger significance due to the fact that in current systems there is a shift from computing oriented to data oriented systems [4]. In order to improve the performance of the whole system, the mismatch between logic and memory has to be decreased. One of the fastest memories on the market is the static random access memory (SRAM). Even though the SRAM as an on-chip memory offers a speed comparable to the logic chip, it could encounter (and is encountering) a scaling problem due to its large cell size (usually six transistors). Further, one of the negative side effects of SRAM scaling is that leakage power of SRAM increases dramatically, constituting one of the main challenges of future systems [5]. On the other side, the DRAM is characterized with smaller memory cell size and higher density at the same technology node. However, its dynamic nature/volatility (need for refreshing every 64 ms) and latency hinder the usage of the DRAM as a memory suitable for the emerging devices. Therefore, an alternative is needed that is:

a) characterized with a small cell size and thus, scalable [1,6],
b) non-volatile [1,6],
c) well matched with the speed of the CPU [1,6],
d) characterized by low power consumption [1,6], and
e) able to store large amount of the data on the defined technology node (price per bit as low as possible) [1,6].

Currently, different emerging concepts (non-volatile random access memories/NV-RAMs) of realization are being considered. Among NV-RAMs, the most researched are:

a) magnetic random access memories (MRAM) based on the spin orientation and the approach of spin transfer torque (STT-MRAM) [6,7,8,9].
b) Resistive random access memories (RRAM) that change the cell's resistivity by 1) changing the filament size; 2) interface switching [6,7,10,11].
c) Phase change memories (PCM) that change their resistivity by altering the layer property between the amorphous and crystalline state [6,12,7].
d) Ferroelectric random access memories (FeRAM) which use the reversal of the ferroelectric's remnant polarization to store binary information [6,7,13,14].

1 INTRODUCTION

Table 1.1 Comparison of emerging non-volatile memories and Flash with volatile DRAM. Values obtained from [15,16]

Parameter/NVM	MRAM	FeRAM	FeFET	PCM	RRAM	Flash NAND	Flash NOR	DRAM
R/W speed	<10 ns	10 ns	10 ns	>100 ns	<10 ns	100μs	1μs-1s	~10 ns
Endurance	1e15	1e15	1e5	>10^8	1e12	1e4	1e6	1e16
Retention	10 years	10 years	10 years	10 years	10 years	10 years	10 years	64ms
Energy	2 pJ/Bit	50 fJ/Bit	< 1 fJ/Bit	3 pJ/Bit	50 pJ/Bit	1 nJ/Bit	10 nJ/Bit	pJ/Bit
Cell size	20 F^2	6 F^2	6-10 F^2	5.5 F^2	4 F^2	4 F^2	10 F^2	6 F^2

In Table 1.1, a comparison of the emerging memories with DRAM and Flash is given. It can be seen that ferroelectric memories offer a low power consumption, ultra-fast operation, retention for more than 10 years and small relative cell size as the main requirements for the future emerging memories. This applies especially to the single-transistor (1T) architecture based on a ferroelectric field-effect transistor (FeFET)[17,18,19]. Due to their remarkable properties both hafnia based FeRAM and FeFET nominate themselves as potential candidates for the NV-RAM that would bridge the speed gap between fast logic and slow off-chip, long term storage. However, both architectures suffer from limited endurance [20,21,17,22]. To address this fundamental question, within this thesis a detailed electrical characterization and comprehensive device modeling of ferroelectric memories was performed.

The dissertation is organized as follows:

Initially, an overview of the properties of the high-k materials used for ferroelectric memories with focus on the material defects is given in Chapter 2. This is followed by the overview of the charge transport within the dielectric materials. Besides charge transport, phenomena of the ferroelectricity are introduced and described based on the two theories: a) Preisach model of hysteresis and b) Landau-Ginzburg-Devonshire (LGD) theory of the phase transitions. The chapter is closed with the overview of the two ferroelectric memory architectures, FeRAM and FeFET, respectively.

Within Chapter 3, fabrication steps and a detailed description of the capacitor and transistor devices under test (DUT) is given. The Chapter 4 briefly covers the dielectric, ferroelectric and reliability characterization methods used within the scope of this thesis.

Chapter 5 is dedicated to the charge transport study and comparison of the dielectrics used in DRAM and FeRAM cells. After the extensive characterization of the dielectric properties, a charge transport model of ZrO_2 based DRAM is developed. Moreover, interface effect and bias fields are studied by exchange of the electrodes on the ZrO_2 devices. The developed model is then extended to the sister oxide HfO_2 that is

1 INTRODUCTION

utilized in novel ferroelectric memories. In addition, reliability improvement tricks from DRAM are applied to FeRAM.

The previously developed charge transport model is expanded based on the detailed characterization in Chapter 6. Field cycling of the ferroelectric thin films is studied with focus on the memory window instability. This results in the development of the 3D grain boundary model of the ferroelectric capacitor (FeCap) that explains life time stages of the ferroelectric memory cell. At the end of the chapter, this FeCap model is extended to ferroelectric metal oxide semiconductor (FeMOS) and FeFET stacks discussing the main design guidelines and strategies needed for future implementation of the device.

Within the final Chapter 7, a solution for the endurance problems of the ferroelectric memories is proposed. Based on the theoretical analysis a device model of new emerging memory is derived. This theoretical analysis together with the model developed pushed the implementation and fabrication of the world first non-volatile, anti-ferroelectric memory.

2 Fundamentals

Within this chapter, the fundamental properties of the dielectric materials with high dielectric permittivity will be reviewed. Focus will be on the binary oxides and two important dielectric materials of the semiconductor industry, HfO_2 and ZrO_2. Due to the fact that these materials are characterized with high defect concentrations that cause increased leakage currents and reliability issues, charge transport mechanism within the dielectric will be discussed. Further, Section 2.3 is devoted to the phase transitions within the high-k materials based on the Landau-Ginzburg-Devonshire theory of phase transitions. Afterwards, ferroelectricity and anti-ferroelectricity in (doped) hafnia and hafnia-zirconia mixtures are discussed. At the end of the chapter two ferroelectric memory architectures, FeRAM and FeFET are covered, respectively.

2.1 High-k Materials

One of the fundamental properties that characterize all dielectric materials is the dielectric constant k. The dielectric constant is a quantity that represents an ability of material to resist the formation of the electric field within it. In contrast to the low-k materials that exhibit negligible change in orientation of molecules under external field, high-k materials undergo a process of polarization that counteracts the externally applied fields.

The most used and studied dielectric material within the semiconductor industry is SiO_2. Offering the thermal stability, a k-value of 3.9 and a band gap 8.9 eV, it represented an irreplaceable ingredient of the transistor stacks. Thus, it became a benchmark for comparison of the dielectrics. Over decades this dielectric material has been successfully scaled and integrated in numerous technology nodes. However, as the consequence of the aggressive scaling and physical thickness reduction under 3 nm, charge carrier flow due to the tunneling (leakage) through the dielectric would become significant and lead to a unacceptable values of circuit power dissipation [23,24,25]. By means of the quantum mechanical tunneling charge carriers are penetrating the ultra-thin SiO_2 creating the defects which finally result in breakdown and failure of the device. Furthermore, this leakage results in severe heating and power consumption of semiconductor device. Therefore, a need for the high-k materials that would provide sufficiently high capacitance while relaxing the thickness scaling became inevitable.

Even though a broad spectrum of high-k materials can be potentially used as dielectric material, these materials need to be able to suppress the gate leakage, while at the same time keeping the gate dielectric capacitance on a high level. Due to the fact that the defects determine the properties of the dielectric, the parameters closely related to the electrical conduction and band engineering of dielectric materials should be considered. The requirements one prospective dielectric material should fulfill, are summarized in the following [26,27].

Electrically or conduction-wise, these films:

- Should possess low defect concentrations, low charge trapping, high breakdown (BD) voltages and generally their reliability should be high.
- Have to have good interface quality, which is directly correlated to carrier mobility that can experience reduction due to scattering on rough interfaces.
- Should be characterized by such energetic gap and barrier heights/conduction band offset (CBO) faced by electrons and holes, that tunneling processes and leakage paths within the dielectric are minimized or ideally prevented completely.
- Should possess a high-enough k-value so that they can be utilized within a reasonable number of the scaling nodes.
- At the end, have to be CMOS integratable. In other words, the replacement dielectric needs to be thermodynamically stable on silicon substrate because it needs to withstand all fabrication steps (e.g to be compatible with thermal budget equivalent of 1000 °C lasting for about 5 s).

An overview of the mainstream high-k materials utilized within the semiconductor and especially memory industry and their most important parameters (such as energetic bandgap, k-value) is given in Table 2.1. In addition, the empirical trends of breakdown field strength, k-value and bandgap correlations are illustrated in Figure 2.1.

Table 2.1 Overview of the main dielectrics utilized in the semiconductor industry and comparison of their k-values and energetic band gap [26,28,29,30] to silicon as reference.

Material	Si	SiO_2	Si_3N_4	Al_2O_3	TiO_2	HfO_2	ZrO_2	Ta_2O_5	$SrTiO_3$
k-value	11.7	3.9	7.4	8.7	15-80	20-40	20-44	25-26	200
Bandgap [eV]	1.17	8.9	5.1	9	3.05	5.9	5.8	4.4	3.2

Based on the requirements listed above and having in mind empirical relations of the k-value dependence on breakdown field ($E_{BD} \sim 1/\sqrt{k}$) and band gap ($E_{BandGap} \sim k^{-0.65}$), any choice of a proper dielectric material represents a tradeoff. Both HfO_2 and ZrO_2 are characterized with high bandgap and consequently high CBO while still providing sufficiently high k-value and thus represent a perfect match and prove themselves to be the optimal solution for semiconductor industry.

2 FUNDAMENTALS

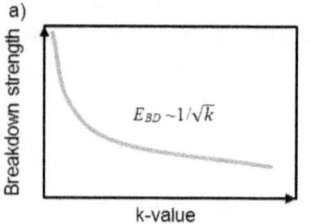 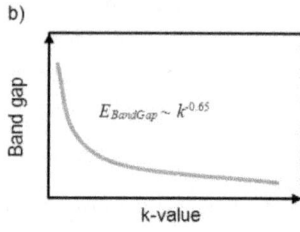

Figure 2.1 Overview of the high-k materials behavior depending on its most important properties such as k-value, band gap and breakdown strength. a) Inverse proportionality of breakdown strength and k-value. b) Inverse proportionality of the band gap and the k-value of the high-k solutions. The trends were calculated based on empirical relations given above.

Due to the very similar k-value and band gap, HfO$_2$ and ZrO$_2$ are considered sister materials [31,32]. In addition, they represent the workhorse and state-of-the art of semiconductor industry. After Intel utilized the amorphous HfO$_2$ as the gate dielectric in 2007, hafnia became standard in the high-k metal gate (HKMG) technology [23]. The following years brought the engineering of the previously amorphous HfO$_2$ and stabilization of the phases with even higher k-values (e.g. tetragonal 25-70 [33,34]) characteristic for the polycrystalline thin films. These films are characterized with a very high oxygen conductivity [35], which proves to be suitable and convenient for the resistive random access memories [36,37,38]. Moreover, the discovery of ferroelectric properties in doped hafnium oxide by Böscke et al.[39] revitalized the research and enabled scaling of the ferroelectric memories. On the other hand, ferroelectric properties within the polycrystalline hafnia closed the gap between the theory and realization of negative-capacitance based steep sub-threshold slope devices [40][41]. Envisioning energy efficiency, those devices are considered to be the holy grail of the modern electronics. Therefore, it can be concluded that polycrystalline hafnia positioned itself as one contender within the HKMG and non-volatile memory industry.

Apart of HfO$_2$ dielectrics, ZrO$_2$ based dielectrics are so far irreplaceable ingredient of the dynamic random access memories stack that are driving the DRAM scaling for already a decade [42,43,44,45,46]. The good compromise of a k-value of around 30 and high band gap enabled suppression of the leakage current and downscaling of the devices. Stacks combining ZrO$_2$ with Al$_2$O$_3$ interlayers enabled steady scaling of the DRAM capacitors to sub-20 nm technology nodes [42,43,44].

Unfortunately, both ZrO$_2$ and HfO$_2$ dielectrics are characterized with high defect concentrations of e.g. oxygen vacancies, which mostly contribute to the leakage through the stack and represent a serious reliability challenge. Based on the defect formation energy, calculations performed by in [24,25] show that the most important defects from the perspective of electrical conduction are neutral (V_o), positive (V_o^+) and

2 FUNDAMENTALS

two-fold positive (V_o^{++}) oxygen vacancies [47,48]. The energetic positions of these inter-band defect states are given in Figure 2.2.

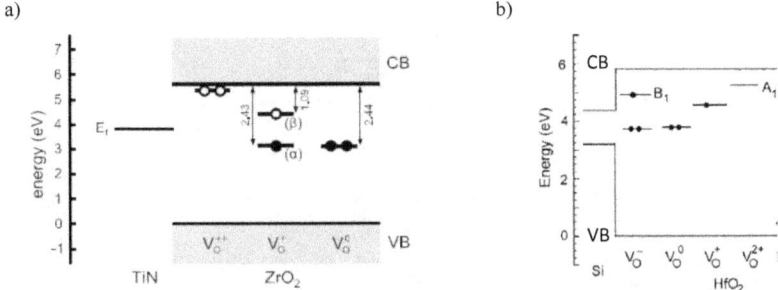

Figure 2.2. Energy level diagram showing the positions of the oxygen vacancy states within the a) tetragonal ZrO_2 and b) HfO_2 band gap. The values of the respective defect energy levels are given in electron volts (eV). For V_o^+ the non-degenerate spin up and down states are denoted (α) and (β), respectively. Open (closed) circles denote unoccupied (occupied) states. CB and VB denote conduction and valance band, respectively. V_o, V_o^+, V_o^{++} (V_o^{2+}) and V_o^- denote neutral, positive, two-fold positive and negative oxygen vacancies. Figures taken from [25,49].

2.2 Charge Transport and Leakage Mechanisms

As mentioned before, further device scaling necessitates the introduction of alternative dielectric materials. Even though the high-*k* materials provided the necessary suppression of the gate leakage and at the same time kept the gate dielectric capacitance on a high level, reliability issues due to dielectric scaling have to be addressed. Due to the lack of the mobile carriers, dielectrics ideally cannot conduct any charge. In spite of the definition of dielectrics as insulating materials and the lack of the mobile carriers, a finite conduction through dielectrics exists. Due to the fact that the performance of the high-*k* dielectrics is mainly affected by the different defects within the crystal lattice, the introduction of high-*k* materials has proven to be one of the most challenging tasks since the beginning of scaling. The charge transport through the dielectric is governed and described by different leakage mechanism. Roughly, those mechanisms can be differentiated based on whether the charge transport is defect/trap mediated or not (see Figure 2.3) into the following:

- without defects (defectless): Schottky emission, direct and Fowler-Nordheim tunneling.
- defect mediated: Poole-Frenkel conduction, trap-assisted tunneling.

Since the suppression of the leakage and purity of the gate stack is essential for the reliable operation of every electronic device and non-memory as well, a detailed analysis and simulation is needed. Therefore, in the following sections an overview of the dielectric conductions mechanisms will be given.

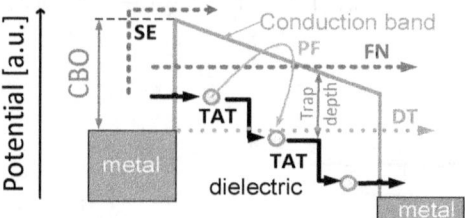

Figure 2.3. Metal-dielectric-metal structure under applied electric field and charge transport mechanism within the film comprising Fowler–Nordheim tunneling (FN), Schottky emission (SE), trap-assisted-tunneling (TAT), Poole-Frenkel emission (PF), direct tunneling (DT). The conduction band offset is denoted with CBO.

2.2.1 Defectless transport mechanisms

Within this section, an overview of the charge-transport mechanisms which are not mediated by the presence of the defects is given. The focus will be on the Schottky emission, direct and Fowler-Nordheim tunneling.

Schottky Emission

Schottky emission represents a field-assisted thermionic emission of an electron over a surface barrier. Namely, a certain number of electrons possess enough energy to overcome energy barrier and are released. The barrier to be overcome is represented through the conduction band offset, CBO which represents energy needed to be given to electron to ionize into the conduction band. Current density J_{SE} governed by Schottky emission [49,50] can be modeled with Richardson equation:

$$J_{SE} = A^{**}T^2 exp\left[\frac{-q\left(CBO - \sqrt{\frac{qE}{4\pi\varepsilon_0 k_r}}\right)}{k_B T}\right] \quad ; \quad A^{**} = \left(\frac{4\pi q m^* k_B^2}{h^3}\right) \quad (2.1)$$

, where A^{**} denotes the effective Richardson constant, while m^*, k_r, ε_0, k_B and h are the effective mass of electron inside the oxide, the dynamic dielectric constant, the vacuum permittivity, Boltzmann's and Planck's constants respectively.

Direct Tunneling

As discussed previously, the scaling of the dielectric is reaching quantum mechanical limits. With the decrease of film thickness to below 4 nm, tunneling effects can be observed. Namely, at high fields dielectric barrier is thin enough to detect the wave/particle duality of the electrons. This conduction mechanism is characteristic for the ultra-thin films on the high fields as well as at low temperatures [49,50].

2 FUNDAMENTALS

The direct tunneling J_{DT} represents a defectless tunneling from the injecting electrode through the dielectric to the opposite electrode and is described by the following equation:

$$J_{DT} = \frac{q^3 q m_e E^2}{8\pi h m^* \left(\sqrt{qCBO} - \sqrt{qCBO - qV}\right)^2} \exp\left\{-\frac{4\sqrt{2m^*}}{3q\hbar E}\left[(qCBO)^{\frac{3}{2}} - (qCBO - qU)^{\frac{3}{2}}\right]\right\} \quad (2.2)$$

, where q is the elementary charge, $\hbar$ reduced Planck's constant, m_e the mass of the electron, m^* the effective mass of the electron, CBO the potential barrier between the metal's Fermie-level and conduction band of the dielectric, and V denotes the applied voltage.

Fowler-Nordheim Tunneling

When applying an electric field across the gate stack, there is a probability for tunneling through the bent barrier. In contrast to direct tunneling, which represents tunneling through a rectangular or trapezoidal barrier from metal to metal, Fowler-Nordheim (FN) tunneling is a tunneling through a triangular barrier from the fermi level of one electrode into the conduction band of the dielectric. The resulting FN, current density J_{FN} can be modeled with the following equation:

$$J_{FN} = \frac{q^3}{8\pi h CBO} E^2 \exp\left(-\frac{4\sqrt{2m^*}\sqrt[3]{CBO}}{3hqE}\right) \quad (2.3)$$

, where q is the elementary charge, h Planck's constant, m^* the effective mass of the electron, CBO the potential barrier between the metal and the dielectric, and the E denotes the applied electric field.

The Fowler-Nordheim tunneling mechanism is the dominating tunneling mechanism in very thin SiO_2 gate dielectrics [49,50]. Additionally, the approximation by Wentzel, Kramers and Brillouin (WKB) [51, 52] is used to calculate the tunneling probability through an arbitrarily shaped barrier.

2.2.2 Defect-mediated transport mechanisms

Within this section, an overview of the defect-mediated transport mechanisms with the focus on the two trap-assisted tunneling (TAT) models is given. The analytical one by Nasyrov [53] and the multiphonon mediated TAT model by Larcher *et al.* [51]. These models will be used in Chapter 5 for detailed analysis of the charge transport within the dielectric of DRAM and FeCAP.

Poole-Frenkel Emission

Poole-Frenkel (PF) emission [49,50,52] can be defined as field-assisted thermal ionization of a carrier from the bulk oxide into the conduction band. Due to interaction with electric field, the potential barrier for emission into the insulator conduction band is lowered. As a result of the thermal fluctuations, an electron is excited from a trap state, which gives it enough energy to jump into the conduction band and exist there until it relaxes into the lower energy state. This conduction mechanism is considered to be bulk-limited. Poole-Frenkel current density J_{PF} is modeled with the following equation:

$$J_{PF} \sim qN_t\mu E \cdot e^{-\frac{q}{k_BT}(W_t - \sqrt{\frac{qE}{4\pi\varepsilon_0\varepsilon_{opt}}})} \qquad (2.4)$$

, where q represents elementary charge, N_t trap density, E applied electric field, k_B Boltzmann's constant, T temperature, W_t trap depth, ε_{opt} represents optical permittivity and ε_0 vacuum permittivity.

However, the model of transport mechanism, given with equation 2.4, does not consider the tunneling thought the potential barrier (dielectric transparency) to the first defect state within the dielectric. Namely, the dielectric transparency can be described with the equation 2.5:

$$P_{In} \sim exp\left(-\frac{2s\sqrt{2m^*W_t}}{\hbar}\right) \qquad (2.5)$$

, where s is the distance between the electron traps, m^* the effective tunneling mass, W_t the trap depth and $\hbar$ the reduced Plank constant.

By calculating emission probability of energetic levels of all defect states and multiplying the rate equation (equation 2.4) with probability for tunneling through the triangular barrier (equation 2.5) a trap assisted Poole-Frenkel mechanism is derived.

Nasyrov multi-phonon trap to trap tunneling model

In 2011 Nasyrov et al. [53] developed and reported a charge transport model based on multi-phonon ionization mediated trap assisted tunneling. In this model, deep defects (electron traps) are represented as single oscillators embedded into the dielectric matrix. The energy dependency of the defect potential from generalized coordinate Q of a system trapped-electron-bound-with-phonon are represented by $U_b(Q)$ terms. Moreover, here the potential energy of the defect-free oscillator has a parabolic shape ($U_b(Q)=Q^2/2$) described with the energy needed for the thermal ionization and the optical ionization of the trap, W_T and W_{opt}, respectively. Details can be found in [53]. The energy configuration for two phonon-coupled traps was obtained from simulations as shown in Figure 2.4. Here, the energetic position "*State I*" represents the energy of a defect that is occupied by a trapped electron whereas in the distance "*s*" (defined as the third root of defect concentration), a second empty defect is located. As defects are represented by/as oscillators that may oscillate, at some point the electron energy becomes coincident with the energy of the other empty defect "*State II*". At this point, the transition v of the trapped electron to the empty state is possible.

2 FUNDAMENTALS

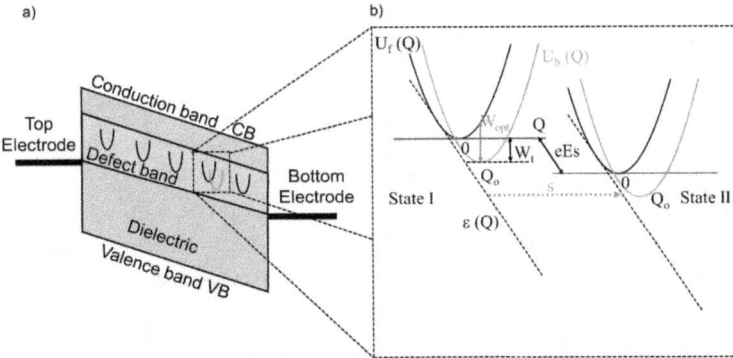

Figure 2.4 a) Sketch of the MIM band diagram with defect states (oscillators) that represent electron traps. b) Configuration diagram for two phonon-coupled traps. $U_f(Q)$ is the potential energy of an empty oscillator (without trapped electron); $U_b(Q)$ is the energy of an occupied oscillator (with trapped electron); "(Q) is the position of the energy level of the trapped electron dependent on coordinate Q. CB and VB denote conduction and valance band, respectively. The red and black lines refer to the initially occupied and empty state, respectively. The most probable tunneling transition for the electron when both oscillators take position $Q_0/2$ is shown by the horizontal dotted arrow. The distance "s" denotes distance between two traps (defects). Figure 2.4 b) simulated based on [53].

In this model a simplified trap-to-trap tunneling probability P_{TAT} is given by:

$$P_{TAT} = 2\frac{\sqrt{\pi}\hbar W_t}{m^*s^2\sqrt{2(W_{opt}-W_t)}} exp\left(-\frac{W_{opt}-W_t}{2k_BT}\right) exp\left(-\frac{2s\sqrt{2m^*W_t}}{\hbar}\right) \sinh\left(\frac{qEs}{2k_BT}\right) \quad (2.6)$$

, where m^* is representing the effective mass of electron, k_B Boltzmann's contant, s the distance between the electron traps, W_{opt} the optical trap ionization energy, W_t the thermal trap ionization energy and E the electric field.

Within the transition rate/probability equation (eq. 2.6), the first exponential term of the equation denotes the activation energy needed to be overcome for the trap ionization. The second term denotes the transparency of the tunnel barrier (tunneling between the states) whereas the *sinh* term represents the dynamical external field dependent modulation.

Accordingly, the current density J_{TAT} can be calculated as:

$$J_{TAT} = qN_t^{\frac{2}{3}}qP_{TAT} \quad (2.7)$$

, where q represents the elementary charge, N_t the defect concentration and P_{TAT} the transition probability described by equation 2.6.

Nasyrov's trap-to-trap tunneling mechanisms is readily available within Sentaurus TCAD [54]. Therefore, TCAD based leakage current simulations together with Nasyrov's analytical TAT model implemented in

MATLAB will be supplemented in Section 5.23 and Chapter 6, with a state-of-the-art charge transport model developed by Larcher et al. [51,38].

Multiphonon Trap Assisted Tunneling

Larcher and co-workers modeled charge transport through the dielectric by developing and applying the statistical multiphonon TAT model, which was proven to accurately reproduce both temperature- and voltage-dependent leakage current through the dielectric [51,55,56]. This model considers that defects assisting the electron transport are positively charged oxygen vacancies located at the grain boundaries (GBs). Further, the model developed by Larcher et al. takes into account both TAT and direct tunneling (DT) contributions [51,55,56]. The electron DT current is computed using the Tsu-Esaki formula [57], while the tunneling probabilities are calculated by WKB approximation.

When calculating the TAT current, each electrically active defect contributes to the charge transport and each independent path depends heavily on the trap with the lowest capture(trap)/emission(de-trap) rate. Within this model, multi-trap percolation paths are automatically accounted for on the basis of the randomly generated defect positions and energies. In order to properly emulate the capture and emission events, the electron-phonon coupling is taken into account. Figure 2.5 illustrates a capture and emission processes. It can be seen that the electron tunneling into a trap is coupled to the release of energy $m\hbar\omega_0$ to the lattice (m is the number of involved phonons, which are assumed to be single frequency ω_0 optical phonons) [51,55,56]. On the other hand, in order to emit electrons from a trap, energy absorption from the lattice is needed. The total current is calculated by computing every capture and emission time constants, accounting for every phonon energy contribution within the dielectric matrix [51,55,56].

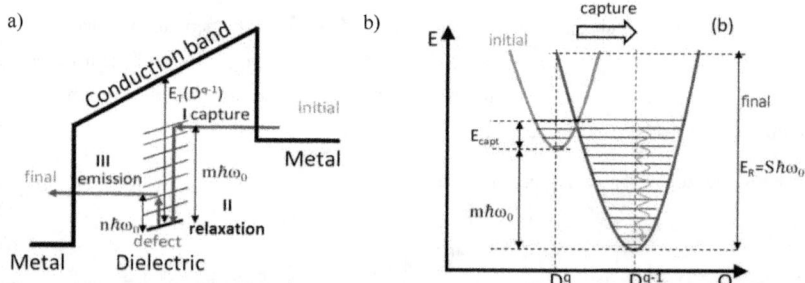

Figure 2.5. a) Sketch of the multiphonon mediated capture, relaxation and emission processes (stages I, II and III, respectively) during the trap-assisted tunneling transport through the dielectric in metal-dielectric-metal structure. b) Energy vs. configuration coordination diagram depicting the energy (i.e., electronic and vibrational) of the system before (red parabola) and after (blue parabola) capture process) of the defect-assisted charge transfer. The red oscillatory line denotes multiphonon mediated relaxation of the electron. Figures reproduced according to the [51,55,56].

2 FUNDAMENTALS

2.3 Phase Transitions and Ferroelectricity within the High-k Materials

Ferroelectric materials are a distinct type of polar oxides. These polar oxides represent a sub-group of the large family of dielectric materials. Dielectric displacement vector D describes the relationship between the electric material and applied electric field and it comprises the vacuum εE and material contribution P:

$$D = \varepsilon_o E + P \qquad (2.8)$$

, where ε_o represents the vacuum permittivity, E the electric field and P the dielectric polarization (density).

The material contribution P is usually linearly dependent on the externally applied electric field. In contrast to traditional dielectric materials ferroelectrics exhibit a nonlinear history dependent polarization change with applied electric field. The observed nonlinearity represents an additional polarization charge called spontaneous polarization P_{sp}. The polarization term P of equation 2.8 can be expanded as following:

$$P = \varepsilon_o \chi E + P_{sp} \qquad (2.9)$$

By combining equations 2.8 and 2.9, the displacement vector D can be expressed as:

$$D = \varepsilon_o (1 + \chi) E + P_{sp} = \varepsilon_o \varepsilon_r E + P_{sp} \qquad (2.10)$$

, where χ denotes the dielectric susceptibility and ε_r the relative dielectric permittivity.

A dielectric crystal which exhibits a spontaneous electric polarization that can be altered between two crystallographically defined states by the application of an external electric field is called a ferroelectric material. Each material that exhibits two clearly distinct equivalent states qualifies itself as a potential memory medium. Therefore, ferroelectrics are a natural choice as information storage materials capable of storing a logical 1 and 0. In contrast to ferroelectrics, a material that exhibits anti-ferroelectric properties is defined as a crystal whose structure can be considered as being composed of two sub-lattices polarized spontaneously in antiparallel directions. According to this theory [58], AFE materials consist of antiparallel dipoles within their crystallographic unit cell which cancel each other at zero field. (see Figure 2.6a). The application of an external field then causes an alignment of the dipoles yielding a positive and negative branch of pinched hysteresis loop. Second theory of anti-ferroelectrics claims that according to the fundamental physical mechanism of AFE materials [59,60], they can be field-forced to undergo a phase transition (see Figure 2.6) into the ferroelectric, polar phase by applying an external electric field, when a critical field value (E_{CR}) is achieved [59,60]. However, the obtained phase is unstable and transforms back into the anti-ferroelectric, non-polar phase when the external field is removed.

2 FUNDAMENTALS

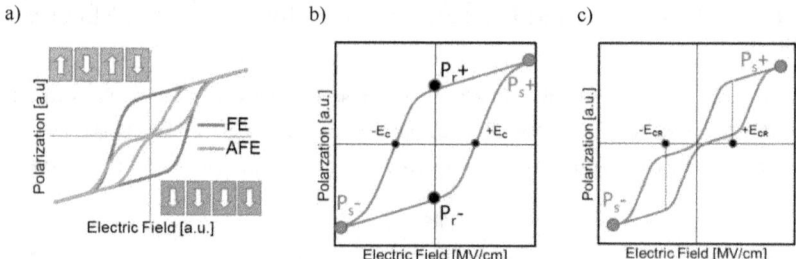

Figure 2.6. a) Domain orientation for ferroelectric and anti-ferroelectric materials at zero fields. Comparison of b) ferroelectric and c) anti-ferroelectric polarization-electric field behavior. E_c denotes the coercive field, P_r remnant polarization and P_s saturation polarization of the ferroelectric, whereas E_{CR} denotes the critical field value of anti-ferroelectric material. Figure b) and c) modified based on Pešić et al. [61]

2.3.1 Theory of the (Anti-)ferroelectric Physics

One of the phenomenological macroscopic descriptions of the anti-/ferroelectric behavior within the polar oxide is given by Landau-Ginzburg-Devonshire (LGD) theory of phase transitions[62][63]. According to the LGD formalism the Helmholtz free energy Φ is:

$$\Phi = \frac{1}{2}\alpha P^2 + \frac{1}{4}\beta P^4 + \frac{1}{6}\gamma P^6 - EP \tag{2.11}$$

, where P represents, ferroelectric polarization, α, β, γ the Landau expansion coefficients and E the electric field.

This series expansion of the Gibbs free energy as a function of the ferroelectric polarization P is an approach to describe the electrical behavior of a ferroelectric in the vicinity of a phase transition [62,63]. For the Landau-Ginzburg-Devonshire equation 2.11 all series powers higher than six are truncated and for symmetry reasons only even powers are included. As such, this expansion series defines the shape of the atomic potential for the switching ion. In the following, a brief discussion on the determination of the Landau coefficients is presented.

First of all, the highest expansion coefficient (γ) has to be larger than zero, otherwise the free energy would approach minus infinity for large values of P, which means the decomposition of the crystal is the most favorable state. Further, simple mathematical transformation of equations 2.9 and 2.11 yields:

$$\frac{\partial E}{\partial P}\left(\frac{\partial \Phi}{\partial P} = 0 = \alpha P + \beta P^3 + \gamma P^5 - E\right)\bigg|_{P=0} = \alpha \iff \alpha^{-1} = \varepsilon_o \chi \equiv \left(\frac{\partial E}{\partial P}\bigg|_{P=0}\right)^{-1} \tag{2.12}$$

Double derivative of the equation 2.11, first with respect to the polarization, followed by another derivation with respect to electric field in the vicinity of the zero polarization yields parameter α which is inversely

2 FUNDAMENTALS

proportional to the electric susceptibility. Thus, it can be determined by measuring the electric susceptibly when the polarization becomes zero. This holds true only when the temperature surpasses the Curie temperature Θ at which the ferroelectric loses its polarization and transforms from the lower symmetry ferroelectric to a higher symmetry non-ferroelectric phase. By expanding α in a series of T around the Curie temperature Θ ($\Theta \leq T_c$ (phase transition temperature)), it can be approximated:

$$\alpha = \alpha_0(T - \Theta) \tag{2.13}$$

The nature of the phase transition is hidden within the parameter β. Landau coefficient β can be extracted experimentally if α_0 and the polarization at zero electric field at a temperature lower than Θ is known.

Based on the parameter nature [62][63] two cases can be distinguished:
- A phase transition of first order occurs when $\beta < 0$ and $\gamma > 0$
- A phase transition of the second order occurs when $\beta > 0$ and $\gamma \approx 0$

In Figure 2.7, both first order and second order phase transitions, are visualized and simulated with regard to polarization dependent free energy.

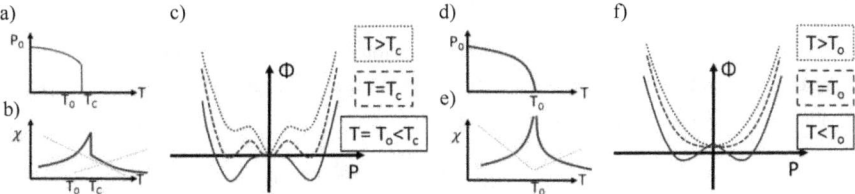

Figure 2.7. Phase transitions according to the Landau-Ginzburg-Devonshire theory. a)/d) polarization and b)/e) susceptibility evolution in the vicinity of the phase transition temperatures T_o and T_c for the first-order phase transition (left) and second order phase transition (right). Temperature dependent Helmholtz free energy vs. polarization in case of c) first order and f) second order phase transition. Figures simulated and reproduced according to [62,63].

2.3.2 Models for Accounting Ferroelectric Properties

Within the scope of this thesis two models for emulation and addressing the ferroelectrics properties of the dielectric materials utilized are:
   a) Landau-Khalatnikov
   b) Preisach based model of the hysteresis.

Landau-Khalatnikov

The core of the Landau-Khalatnikov model is based on the previous discussion on the Landau-Ginzburg-Devonshire theory of phase transitions of ferroelectric materials[62,63]. In addition to equation 2.11, damping factors are added to the theory gaining the ability to model the transient behavior of the ferroelectric behavior. Similar to the model presented in [40], the model was implemented within the MATLAB software package comprising multiple grains.

Preisach Model of the Hysteresis

Hysteretic behavior of ferroelectric and ferromagnetic materials is most commonly represented using mathematical, Preisach model of hysteresis. It expresses the macroscopic hysteresis as the sum of microscopic bi-stable units called hysterons. Hysterons can be associated with ferroelectric switching of a single domain. Main parameters are the electric field needed for switching (E_s) to the positive polarization $+P$ and the electric field needed for backswitching (E_{bs}) to the negative polarization $-P$.

A Preisach based, numerically stable model of the hysteresis [64] is readily available within the Synopsys TCAD package used for device simulations called Sentaurus device [54]. Due to the fact that Preisach hysteresis model within the TCAD Sentaurus device is slightly modified with respect to the original one, its derivation will be discussed in the following paragraphs.

As mentioned in the beginning of this Chapter, the solution of the Gauss's law (3rd Maxwell equation) represents dielectric displacement vector, D that comprises the linear, εE and nonlinear part, P. Nonlinearity, exhibited by polarization charge, is described by:

$$P = c \cdot f(c, E \pm E_c) + P_{off} \qquad (2.14)$$

In equation 2.14, $f(c, E \pm E_c)$ represents the shape function employed for calculation of the non-saturated polarization switching (subcycles). The shape function implemented within the Sentaurus device is a hyperbolic tangent. Therefore, the hysteresis is modeled as:

$$P = c \cdot P_s \cdot \tanh\left[\frac{1}{2E_c} \ln\frac{P_s+P_r}{P_s-P_r}(E \pm E_c)\right] + P_{off} \qquad (2.15)$$

, where c represents the proportionality constant, P_s the saturation polarization, P_r the remnant polarization, P_{off} the offset polarization, E the electric field and E_c the coercive field.

The model described by equation 2.15 was extended by adding the transient properties, where effective polarization $P_{eff}(t)$ is a function of the effective, transient electric field:

$$P_{eff}(t) = f(E(t)) \qquad (2.16)$$

2 FUNDAMENTALS

For the calculation of the effective field a transient parameter was added to a static electric field E_{static} component:

$$E(t) = E_{static} + \tau_{eff}\frac{dE}{dt} \qquad (2.17)$$

, where τ_{eff} represents a material dependent time constant[54, 64].

Furthermore, the transient polarization term was added:

$$P_{pol} = -\tau_p\frac{dP}{dt} \qquad (2.18)$$

, where τ_p represents material specific time constant [54,64].

The introduction of this transient term (P_{pol}) can be physically interpreted as an increase of the coercive field which results in flattening of the hysteresis. However, such system can emulate only a limited number of frequencies. Therefore, to broaden the frequency range of the model a material nonlinearity P_{nonlin} was included ($P=P_{pol}+P_{nonlin}$). The material nonlinearity P_{nonlin} is modeled with:

$$P_{nonlin} = ck_{nonlin}(P - P_{eff})\frac{dE(t)}{dt} \qquad (2.19)$$

, where c and k_{nonlin} represent a material dependent proportionality constants [54,64].

Within the scope of this thesis, the time constants of Preisach model were set to zero (instantaneous switching), thus equation 2.17 simplifies to equation 2.15. The capability of both Landau-Khalatnikov and Preisach models is shown in the Figure 2.8 where a full range fit was obtained. In contrast to the Preisach model which comprises averaging effect through the sum of microscopic bi-stable units called hysterons yielding a smooth transition of polarization reversal, Landau-Khalatnikov model is characterized with abrupt polarization reversal. Therefore, the smoothing of the polarization change and averaging effect is obtained by utilizing a high number of domains (e.g. 1000 domains in simulation depicted in Figure 2.8a) that have Gaussian distribution of determining parameters (internal bias (μ=0.12 V, σ=0.1 V) and landau coefficients $\alpha(\mu$=-4.8·10^8 m/F, σ=4.7·10^7 m/F), $\beta(\mu$=6.8 m^5/F·C^2, σ=1.76 m^5/F·C^2), $\gamma(\mu$=0 m^9/F·C^4, σ=1 m^9/F·C^4)).

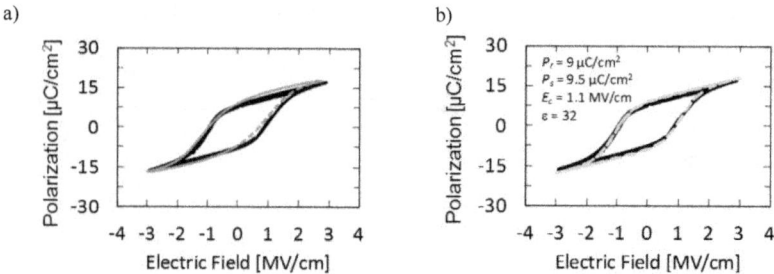

Figure 2.8. Capability of the emulation of the ferroelectric properties and simulation of the history dependent polarization using a) Landau-Khalatnikov and b) Preisach model of the hysteresis available within the Sentaurus device. Figure b) taken from [65].

2.4 Ferroelectricity in HfO_2

In contrast to the traditional perovskite based ferroelectrics like barium titanate (BTO), lead titanate (PTO) and lead zirconate titanate (PZT), doped HfO_2 based ferroelectrics are rather novel ferroelectric-like materials. Since its discovery by Böscke and co-workers [39], they draw significant attention due to the fact that they bridged the gap between the PZT-based ferroelectric devices (90 nm technology node [66]) and the state of the art CMOS technology (1x and 2x nm node). Doped HfO_2 reveals full scalability and compatibility with CMOS-based electronics enabling scaling of the ferroelectric field effect transistors FeFET to 28 nm technology node based on standard HKMG technology [18,17].

The history behind the discovery of the ferroelectric properties within the HfO_2 is based on the aim of the semiconductor industry to stabilize a high-k tetragonal phase within the doped HfO_2 [67,68,67] that was potentially considered to be the most suitable material for the novel DRAM nodes. Because more charge can be stored at the same voltage, this increased permittivity would be beneficial for the read-out of the DRAM cells. Incorporating different concentrations of the Si dopant within the hafnia, together with consecutive anneal leads to an anomalous behavior of the polarization-voltage characteristics of the hafnium oxide. [39] Up to 3 cat% of Si content, the expected classic linear P-V characteristics, i.e. paraelectric behavior was observed. However, with the increase of the Si content to 4.4 cat%, the P-V curve exhibited a nonlinear behavior characterized by a hysteretic characteristic as it is commonly observed for ferroelectric materials. A further increase of the Si content led to a double butterfly curve and pinched hysteretic polarization-voltage relation typical for the anti-ferroelectric like materials.

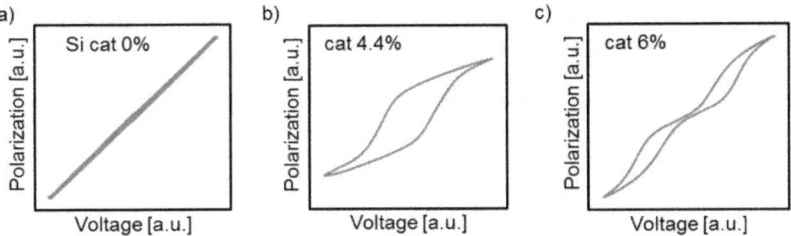

Figure 2.9 Doped HfO$_2$ response on the externally applied electric field., Polarization-voltage characteristics for a) paraelectric b) ferroelectric and c) anti-ferroelectric hafnium oxide.

Based on the X-ray diffraction (XRD) structural analysis, Böscke and co-workers [39] assumed that the non-centrosymmetric orthorhombic Pca2$_1$ phase is responsible for the ferroelectric-like behavior within doped hafnia. On the other hand, cubic and monoclinic phases are correlated with paraelectric behavior, whereas the tetragonal phase is considered to be responsible for the observed anti-ferroelectric-like properties. Simulations of the different phase unit cell of the hafnia performed by Materlik *et al.* [69,70] are shown in Figure 2.10. A monoclinic phase of space group P2$_1$/c is found in bulk hafnia at room temperature. Increasing temperature to around 1700 °C, a tetragonal phase of space group P4$_2$/nmc can be stabilized. Upon a temperature to around 2600 °C hafnia transforms into a cubic Fm3m phase.

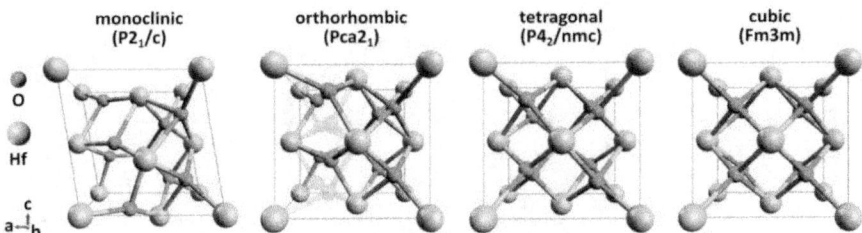

Figure 2.10. Simulated stabilized unit cells of hafnia: paraelectric cubic Fm3m, tetragonal P4$_2$/nmc and monoclincic P2$_1$/c phases, ferroelectric orthorhombic Pca2$_1$ phase. Figure taken from [69][70].

Due to the fact that diffractograms are rich in overlapping peaks, the identification of the Pca2$_1$ space group of orthorhombic phase by XRD is proven not to be most suitable [39,70]. Moreover, isolation of the certain phase is additionally hindered by the polycrystalline nature of the doped hafnia favoring a coexistence and mixture of different phases. Just recently, a research group from North Carolina State University reported the existence of the orthorhombic Pca2$_1$ phase to be responsible for the ferroelectric behavior within the Gd

2 FUNDAMENTALS

doped hafnia.[71] Combining scanning transmission electron microscopy (STEM) with electron diffraction the originally proposed orthorhombic phase could be isolated.

Following the original Si doped HfO_2, various dopants were tested in order to tailor the ferroelectric properties within the HfO_2. Up to now, the ferroelectricity has been proven for various HfO_2 dopants that can be differentiated by valence, ionic radius, crystal radius with respect to the host hafnium atoms. A brief overview is given with Table 2.2 and Figure 2.11.

Table 2.2 Various dopant atom parameters with respect to the properties of the host Hf atoms [72]:

Dopant	smaller then host		Host-reference	Hf-like	larger then host			
	Si	Al	Hf	Zr	Y	Gd	La	Sr
Valence Charge	4+	3+	4+	4+	3+	3+	3+	2+
Crystal radius[pm]	-	-	0.97	0.98	1.159	1.193	1.3	1.4
Ionic radius[pm]	-	-	0.83	0.84	1.019	1.053	1.16	1.26

From Figure 2.11c, a correlation between coercive field and atomic radius can be deduced. Exceptions from these trends are host-like Zr which cannot be exactly considered as the dopant but rather intermixing material (1:1 ratio with host), and the La dopant which yielded one of the lowest E_c.

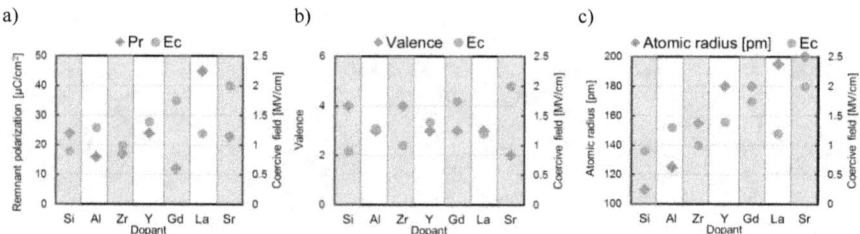

Figure 2.11. Ferroelectric film determining parameters dependence on the dopant atom. Correlation of coercive field with a) remnant polarization b) valence and c) atomic radius of the dopant atom[18][73].

A list of dopants and its concentration within the host material can be utilized in order to tailor the property of the material by stabilizing different phases [39]. In Figure 2.12a various dopants and material properties are given with respect to the amount of dopant within the film. Silicon is one of the best examples according to which can be seen how the doping influences stabilization of all: paraelectric, ferroelectric and anti-ferroelectric properties (see Figure 2.9). It can be seen that just slightly Si-doped HfO_2 films exhibit paraelectric behavior. An increase of the Si doping within the film ferroelectric properties are obtained.

2 FUNDAMENTALS

Further doping results in anti-ferroelectric behavior characterized by a pinched hysteresis loop. Finally, it can be seen that an increase of doping concentration beyond 6 cat% results in paraelectric characteristics again.

Interestingly, the performance of the stack in terms of endurance strongly depends on the phase/doping/dielectric behavior. One of the most interesting properties of the AFE films is that they possess a high endurance [74,75] strength. From Figure 2.12b can be seen that films with AFE properties exhibit an at least three orders of magnitude higher endurance with respect to their FE counterparts.

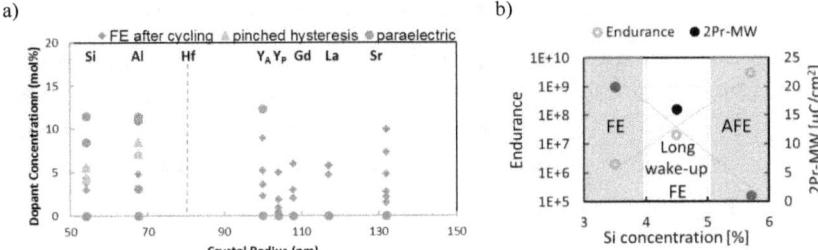

Figure 2.12. a) Dopant concentration induced phase transitions. b) Endurance and $2P_r$ as a function of Si content in Si:HfO$_2$. AFE material exhibits a better endurance compared to FE materials. Endurance was recorded at ±3.5 V and 300 kHz. Figure a) taken from [76] and b) from study by Pešić et al. [77].

2.5 Ferroelectric Memories

Each material which as the fundamental property possesses two clearly distinct states (that can be altered) nominates itself as a potential medium for the building of the memory. Accordingly, ferroelectrics promote themselves as information storage materials capable of storing logical 1 and 0. Within the ferroelectric memory devices, an external field is utilized to alter the state of the ferroelectric domains, consequently changing the polarization state of the material. As discussed above, ferroelectrics are capable of preserving the distinct memory state without presence of an external excitation. This feature classifies the ferroelectrics into the non-volatile materials.

2.5.1 Ferroelectric Memory History and Architecture Overview

The first version of the ferroelectric based memory was proposed already in 1950s [78,79]. Barium titanate crystals (BTO) were the base material of this memory that was realized in matrix/cross-point architecture. However, this concept suffered from disturb effects on the neighboring cells, that were purely controllable in the chosen architecture. It did not take long before two additional ferroelectric memory architectures were considered: Ferroelectric random access memory FeRAM and ferroelectric field effect transistor FeFET) [78,79]. Ramtron was the first company that introduced the mass production of the ferroelectric memories in 1993 [80]. More than a decade later the most aggressively scaled technology node was commercialized by Texas Instruments [81,66,82]. These memories are based on the PZT, which is one of the most prominent ferroelectrics, but unfortunately not the best material with respect to scaling and compatibility with CMOS processing. Thus, the miniaturization of PZT based device ended at the 90 nm node [81]. Besides PZT, strontium-bismuth-tantalate (SBT), bismuth-ferrite (BiFeO$_3$) and polymer based polyvinylidene fluoride (PVDF) mixture with tetrafluoeroethylene (TFE) were examined. Despite the huge efforts invested, none of these materials reached commercialization by entering mass production.

With the discovery of the ferroelectricity in doped HfO$_2$ thin films, the gap between ferroelectric memories and state-of the-art CMOS technology nodes was bridged [46,45,83,84,85]. Utilizing Si:HfO$_2$, the world's smallest FeFET was realized in 28 nm HKMG technology. In the following, two most promising architectures for the implementation and operation of ferroelectric memories will be addressed.

Ferroelectric Random Access Memory (1T-1C Architecture)

The concept of the ferroelectric random access memory is one of the basic non-volatile memory cell architectures. As in the DRAM case, a memory cell consists of one access transistor and one storage capacitor (1T-1C) (see Figure 2.13a). In contrast to DRAM, where the storage element is a capacitor with a linear dielectric, FeRAM utilizes a non-linear, ferroelectric capacitor. Exactly this ferroelectric based storage capacitor grants the non-volatility to the FeRAM in contrast to the volatile DRAM. In addition to this fundamental difference, the plate line is not case grounded, as in the DRAM, but plays significant role in the operation of the cell. During the read operation, to access the capacitor a word line (WL) is pulsed, while simultaneously a voltage pulse is applied at the plate line (to drive a ferroelectric within the capacitor in saturation) (Figure 2.13b-c). This pulse results in the capacitor ending up in always one and the same polarization state. Depending on the polarization state before the pulse, which represents a stored 0 or 1, the resulting transient current either charges the bit line capacitance or leaves it at around 0 V. Similar to DRAM, this operation is destructive (destructive read-out, DRO) and the polarization state of the storage element after the readout corresponds to the polarity of the read voltage. Thus, the readout operation comprises a re-writing step of the memory cell to its initial state. The most important feature of the DRAM is the storage capacitance, whereas the figure of merit of the FeRAM is the remnant polarization, which defines the difference in charge flowing to the bit line between the two memory states 1 and 0. Even though FeRAM and DRAM possess, on the first glance, two different figures of merit, the storage charge is in the basis of both. Therefore, with scaling and consequent reduction of the device area, high remnant polarization is needed to secure a stable read operation.

2 FUNDAMENTALS

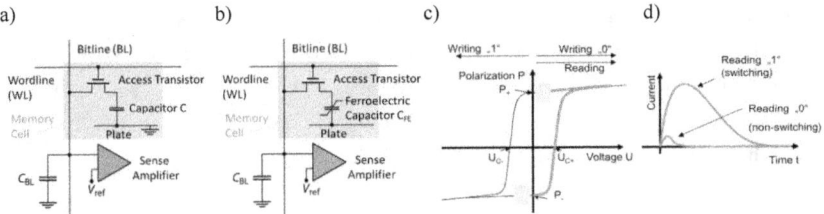

Figure 2.13: Schematic layouts and configuration of a) standard DRAM cell and b) FeRAM cell. c) and d) readout operation of the FeRAM depending on the previous state of the memory. Figures a), b) and c), d) reproduced according to [86] and [87], respectively.

Ferroelectric Field Effect Transistor (1T Architecture)

Similar to the miniaturization of the volatile DRAM, also FeRAM scaling suffers from the decreasing amount of charge which is coupled to the cell size and resulting capacitor area reduction. In addition to the discrete cell size, the second drawback is the destructive readout and the resulting need for a re-write operation of the stored polarization state. A second ferroelectric memory concept, the one transistor (1T) cell architecture, offers a combination of the small cell size and non-destructive read operation. Here, the ferroelectric is directly integrated into the gate stack of the field effect transistor, by being deposited atop of an interface buffer layer. The basics of the operation of this cell effect of the polarization state on the conductivity of the transistor channel (depletion or accumulation). This results in two different threshold voltages V_{th} which denote logical 1 and 0.

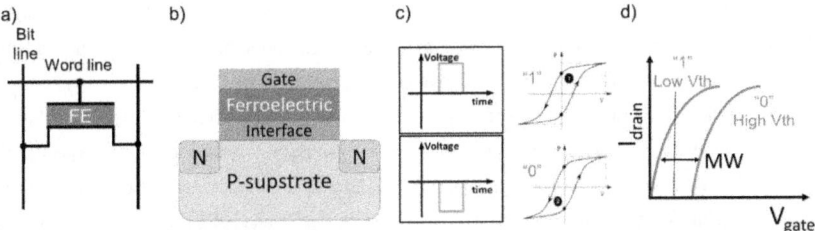

Figure 2.14 a) Schematic of a FeFET cell in NAND configuration. b) MFIS based configuration of FeFET c) program (top) and erase (bottom) state of the FeFET together with respective position 1 and 2 on the hysteresis curve induced with corresponding applied pulses. d) Transfer characteristic of the FeFET with memory window as the voltage difference between two clearly separated states caused by two different polarization states.

Binary information is written by applying a gate voltage via the word line in order to switch the polarization by overcoming the coercive field of the ferroelectric film. The readout of the device works without switching or disturbing the polarization state (non-destructive readout; NDRO) and is performed by applying a source-drain voltage via the bit line. At the same time, a fixed voltage is applied via the WL on the gate of the device in the middle of the MW, which is defined as the difference of the low and high V_{th} state. Finally, either a high or a low drain current is sensed at the bit line. It is important to note that the V_{th} state changes counter clock wise (CCW) when the gate voltage is swept from negative to positive voltage and back, which results in transition from low to high V_{th} state and back.

2.5.2 Hafnium-Based Ferroelectric Memories

The state-of-the-art 1T-1C FeRAMs use PZT-based planar capacitor as a storage element realized in 90 nm technology, thus enabling only a low-density memory. However, the integration of high-density FeRAM cells requires the replacement of the planar two-dimensional (2D) capacitors with three-dimensional (3D) capacitors that would ensure a sufficient capacitance on a limited cell area. Even though the recent study [88] reported the integration of 3D PZT capacitors with a comparable performance to the planar devices, usage of heavy metals within the CMOS lines is undesired due to the disposal and processing issues [89]. Besides, a thickness reduction of the PZT might results in increasing relevance of dead layer regions and is questionable with respect to integration to sub 20 nm nodes. Hafnia, as a standard material within the semiconductor industry eases a replacement of PZT as the storage medium in ferroelectric memories. Concerning the 1T-1C cell architecture, a good example for the capabilities of ferroelectric hafnia is the 3D integration of the Al doped hafnia into high aspect ratio trench capacitor structures [90]. Moreover, doped hafnia as a fully CMOS compatible material enabled the scaling of the 1T architectures and realization of the FeFET device in 28 nm HKMG technology node [91,92,65]. The simplicity of the fabrication and CMOS compatibility is reflected through the fact that ferroelectric (orthorhombic) phase in doped hafnia was stabilized during the source/drain implant activation step, by a flash anneal processing step at 1000 °C. Beside the scalability and full CMOS compatibility, excellent retention properties were reported [46,92]. Even though the retention of presented memory architectures can be extrapolated to a 10-year specification target, both FeRAM and FeFET memory concept suffer of rather limited endurance as the second important criterion for a memory. Typical endurance characteristics of the FeCap and FeFET are given in Figure 2.15a and 2.15b, respectively. It can be seen that the endurance of both devices is characterized by an initial increase of the memory window which is followed by the fatigue of the device. In contrast to the FeRAM case, where hard breakdown occurs before the closure of the memory window, field cycling of the FeFET results in complete closure of the memory window. The classical transistor characteristics remain measurable, which means that the device still works as a transistor, but its memory properties got lost.

2 FUNDAMENTALS

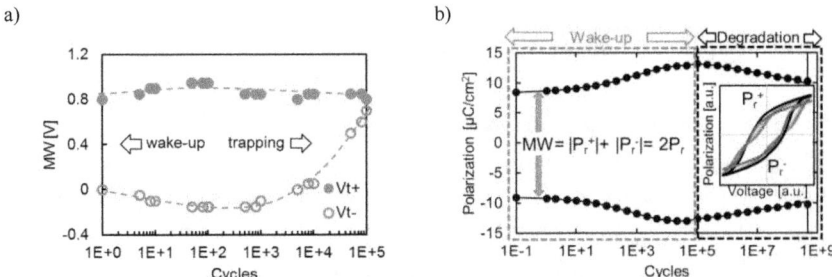

Figure 2.15. Endurance characteristics of the a) FeFET measured on the multi-structure with program erase condition of -5/4.5 V. b) FeCap recorded at 10 kHz at 100 kHz at 3.3V.

To address the question of the root cause of this degradation, a detailed study of field cycling evolution of these two polarization based memories will be presented. This study is structured as follows:

- A description and fabrication details of the test vehicles used within the scope of this thesis are given in the Chapter 3.
- An overview of characterization methods utilized in this work is given in the Chapter 4. These include electrical measurements identifying the dielectric, as well as memory performance.
- First, a detailed analysis of the charge transport and dielectric strength will be addressed. In order to rule out any influences of the ferroelectric switching in the Chapter 5 a detailed screening of the charge transport and interface influence on the dielectric behavior will be presented. Further, at the end of the charge transport section, the influence of different electrodes on the internal bias on the transport characteristics will be addressed. Finally, the charge transport study will be extended to include the ferroelectric high-k (Sr:HfO$_2$) materials.
- After the interface study, a comprehensive electrical characterization and modeling of the wake-up and fatigue behavior within the Sr doped hafnia is presented in Chapter 6. Besides the model describing the physical mechanisms behind the field cycling and limited endurance of the hafnia based ferroelectrics is proposed.
- The final Chapter, is dedicated to the possible improvements as well as with the proposal of the concept of non-volatile anti-ferroelectric random access memory (AFE-RAM). There merits of the novel non-volatile memory concept will be defined.

3 Description of the Devices Under Test

Within the scope of this work, various gate stacks were fabricated and tested. Different capacitor structures mimicking different gate stack were tested in order to study the certain performance, responsible mechanism and finally derive conclusions about optimized stacks that would enable longterm-stable non-volatile memories. Among the versatile stack tested within this work, the focus was on:

a) Metal-insulator-metal (MIM) structures comprising TiN bottom electrode and different top electrodes (TiN/Ru/RuO$_2$/Pt);

b) Above listed metal electrodes were sandwiching the dielectric stack comprising pure ZrO$_2$, HfO$_2$ or ZrO$_2$/Al$_2$O$_3$/ZrO$_2$ (ZAZ) thin films;

c) Dielectric stack equivalents fabricated on top of a SiO$_2$ interface buffer layer and silicon substrate form metal-ferroelectric-insulator-semiconductor (MFIS) stacks used as the transitional study medium between a MIM and FeFET device;

d) Ultra-scaled, fully CMOS compatible FeFET multi-structures integrated in 28 nm HKMG technology.

The bases of each device stack as well as the detailed fabrication procedures are listed in following paragraphs.

3.1 Metal-Insulator-Metal (Semiconductor) Capacitor Structure

3.1.1 ZrO$_2$ Based Capacitors

In this part, metal-insulator-metal structures were deposited on a silicon substrate in a three-chamber ultra-high vacuum (UHV) physical vapor deposition (PVD) cluster tool produced by BESTEC. The full MIM capacitor stack was fabricated without breaking the ultra-high vacuum, which avoids film and interface contamination. The TiN bottom electrode was deposited by PVD on a p-type Si substrate. Further, the deposition of the ZrO$_2$ dielectric stacks was performed at two different chamber pressures (0.001 and 0.005 mbar) at 100 °C. Independent of the deposition chamber pressure, the ZrO$_2$ stacks were crystallized by an 800 °C anneal for 20 s in a nitrogen atmosphere. Top electrodes (TE) were deposited in the same PVD tool (TiN) and BESTEC e-beam evaporation tool (Ti), respectively. To form individual TiN top electrode structures, platinum dots were evaporated onto the top Ti layer as a hard mask. Then, the top TiN layer was removed around the pads with a diluted standard clean 1 (SC-1) solution consisting of H$_2$O, H$_2$O$_2$ and NH$_4$OH in a ratio of 50:2:1. The physical thickness of ZrO$_2$ deposited at lower pressure was about 16 nm,

3 DESCRIPTION OF THE DEVICES UNDER TEST

whereas the deposition at higher pressure yielded in a 9.6 nm thin dielectric stack. The thickness was measured by X-ray reflectivity (XRR) on a Bruker D8 Discover with Cu-K$_\alpha$ radiation (0.154 nm wavelength).

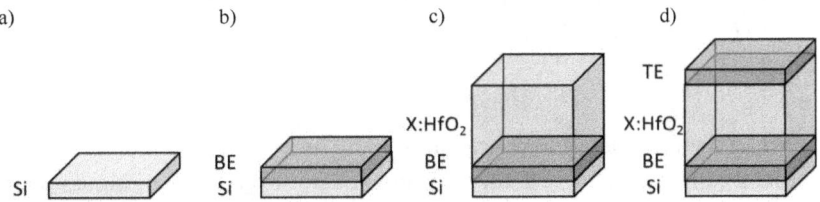

Figure 3.1 Process flow used for fabrication of the gate stacks under test. a) Si substrate, (b) PVD deposited bottom electrode, c) ALD deposited dielectric, d) PVD deposited top electrode.

For comparison, ZrO$_2$ layers were fabricated by atomic layer deposition (ALD) in an Oxford OPAL system on PVD-deposited TiN bottom electrodes. Both top and bottom electrodes were deposited at room temperature whereas the dielectric stacks were processed using trimethylaluminum (TMA), a Zr based metal organic precursor (TEMAZr) and H$_2$O as reactants at 230 °C. After capping the 12 nm thin dielectrics with a TiN top electrode, all samples were annealed for 20 s at 500 or 800 °C in N$_2$ atmosphere.

To explore the benefits of 3D integration and related reliability influences, 7.5 nm thin ZAZ layers were deposited by ALD on chemical vapor deposition (CVD) based TiN bottom electrodes by using TiCl$_4$ and NH$_3$ precursors. Afterwards, the dielectric stack was formed using trimethylaluminum (TMA), a Zr based metal organic precursor (TEMAZr) and O$_3$ as reactants. The stack was crystallized during the top electrode TiN deposition process at 450 °C for 15 min. Details of the structure can be found in [93].

The corresponding MIS structures were processed identically but without bottom electrode directly on the silicon substrate without removing the native oxide. Figure 3.2a-d shows detailed sketched of the film stacks used in this work.

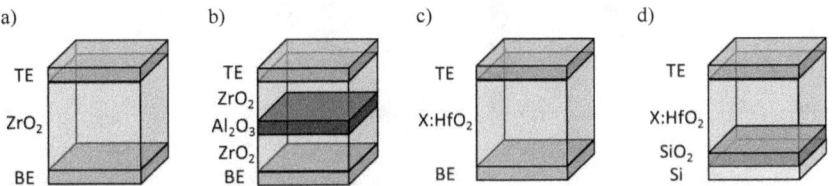

Figure 3.2 Devices under test. a) TiN/ZrO$_2$/TiN, b) TiN/ZrO$_2$/Al$_2$O$_3$/ZrO$_2$/TiN, c) TiN/X:HfO$_2$/TiN and d) TiN/X:HfO$_2$/SiO$_2$/Si capacitor structure used as the test vehicles within this thesis.

3 DESCRIPTION OF THE DEVICES UNDER TEST

In order to measure the workfunction of the different electrodes terraced oxide structures were prepared. Thermally grown SiO_2 on a Si sample was etched in diluted hydrofluoric (HF) to form terraces of different thickness. Thermal oxidation was followed by 10 minutes long forming gas processing step in H_2 atmosphere at 400 °C. Afterwards the ZrO_2 high-k material and TiN electrode were deposited. Structuring and creation of the discrete $Si/SiO_2/ZrO_2/TiN$ MOS structures were performed using the SC1 solution.

3.1.2 Doped HfO$_2$ Based Capacitors

The TiN/Sr:HfO$_2$/TiN metal-ferroelectric-metal (MFM) stacks were fabricated on Si substrates. A 10 nm TiN bottom electrode and the 10 nm Sr:HfO$_2$ dielectric were deposited by reactive sputtering at room temperature and atomic layer deposition at 300 °C, respectively. The Sr dopant content was determined as reported earlier [94]. Further sample preparation included the reactive sputtering of the 12 nm TiN top electrodes at 200 °C. After completion of the stack, Sr doped HfO$_2$ was crystalized by a 20 s long anneal at 800 °C in nitrogen atmosphere. Finally, dots consisting of 10 nm Ti (adhesion layer) and 50 nm Pt on top were deposited in an electron beam evaporator using a shadow mask defining the lateral size of device of 9500 μm². An overview of the dopants used is listed in Section 2.4.

3.2 Ferroelectric Field Effect Transistor (FeFET)

Si:HfO$_2$-based metal-ferroelectric-insulator-semiconductor field effect transistors (MFIS-FETs) with a poly-Si/TiN/Si:HfO$_2$/interface buffer oxide/Si gate stack were processed using state-of-the-art 28 nm high-k metal-gate technology [65] on 300 mm industrial manufacturing process line at GLOBALFOUNDRIES Dresden Module One LLC & Co. KG. The gate stack consists of 1.2 nm interfacial SiON layer (obtained with plasma nitridation of the chemical oxide), 9 nm thick ALD-deposited ferroelectric HfO$_2$ layer doped with 4.4 mol% SiO$_2$ and 8 nm thick PVD-deposited TiN metal layer. Spike anneal at 1050 °C was the maximum thermal budged experienced by the stack. This thermal budged was sufficient for crystallization of the ferroelectric high-k layer.

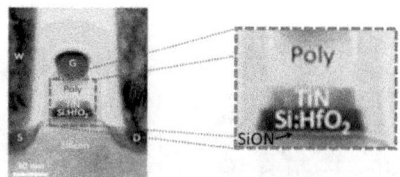

Figure 3.3 TEM micrograph of 28 nm Si:HfO$_2$-based FeFET with an enlarged image of the gate stack. Process flow used for fabrication of the gate stacks under test is following: 1.2 nm thick SiON interfacial buffer layer obtained by nitridation of the chemical oxide; 9 nm thick ALD-deposited Si doped HfO$_2$ layer; PVD deposited 8 nm thick TiN metal gate electrode; Crystallization of the ferroelectric high-k was induced by spike anneal at 1050 °C.

4 Characterization Methods

Due to the large number of the experimental methods utilized for the within the scope of this thesis, this chapter gives an overview of main characterization methods and techniques. The supplementing characterization methods that were used to complement these main tests will be described during the discussion of the results in Chapters 5 and 6. The applied characterization methods can be differentiated as follows:

a) Dielectric characterization methods: current-voltage, capacitance-voltage and leakage current defect spectroscopy (LCDS) will be discussed with focus on the reliability measurements such as dielectric relaxation (DA), breakdown voltage tests (VBD).
b) Ferroelectric characterization methods comprising PUND, polarization-voltage, and first order reversal curves (FORC)
c) Memory characterization methods comprising endurance test, retention test.
d) Physical characterization methods comprising scanning electron microscopy (SEM), transmission electron microscopy (TEM) and X-ray diffraction measurements.

4.1 Dielectric Characterization Methods

4.1.1 Leakage Current Defect Spectroscopy

Besides the typical dielectric characterization methods such as *I-V* and *C-V* test leakage current defect spectroscopy is one of the most powerful techniques for dielectric quality characterization. This method was reported for low-*k* materials by Gishia et al.[96], whereas Pešić et al.[43] adjusted it for the application to high-*k* materials. The basis of the method reported in [97] comprises two consecutive *I-V* sweeps. After the initial sweep a certain number of electrons remains trapped within the dielectric which consequently modifies the electrostatics within the film as well the dielectric transparency. Accordingly, the second *I-V* characteristics is shifted by voltage ΔV. Based on the capacitive principle, the voltage shift ΔV is directly proportional to the total amount of trapped charges:

$$\Delta V = \frac{d}{2} \cdot \frac{-Q_{traped}}{\varepsilon_0 k_{ox}} \qquad (4.1)$$

, where ε_o represents the relative permittivity, k_{ox} the dielectric constant of the material, q the elementary charge, ΔV the field shift between the *I-V* sweeps, Q_{traped} the trapped charge within the dielectric and d the thickness of the dielectric.

4 CHARACTERIZATION METHODS

The measurement procedure utilized within this work consists of three steps, during which the transient current was monitored (See Figure 4.1). The initial voltage sweep step was followed by a constant voltage stress step. At the point in time when the current drop became negligible, which is a result of the filling of the defect states, the sweep was continued until dielectric breakdown. In the following voltage sweep, an additional voltage was needed to reach the current level as measured before the stressing step. By using the voltage difference and assuming uniform density of charges (equation 4.1) in dielectric the trap density N_t can be calculated as:

$$N_t = \frac{2 \Delta V \, \varepsilon_0 k_{ox}}{q \, d} \Big|_{Q_{traped} = qN_t} \quad (4.2)$$

, where ε_o represents the relative permittivity, k_{ox} the dielectric constant of the material, q the elementary charge, ΔV denotes a voltage needed to be applied to the system in order for the current to reach the magnitude previous to the electron trapping (CVS stage) and d the thickness of the dielectric.

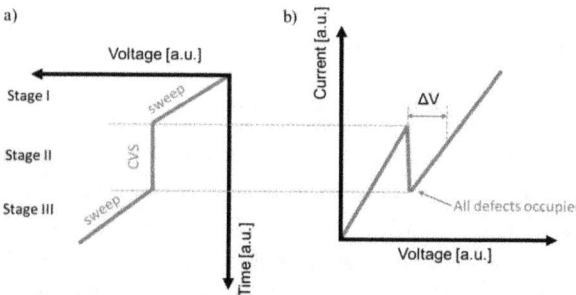

Figure 4.1 a) Leakage current defect spectroscopy measurement procedure depicting voltage transient comprising three stages: (I) sweep, (II) CVS-stress/trapping and (III) continuation of the sweep. b) Resulting I-V and enlarged stage two indicating current drop due to the charge trapping in Stage II. ΔV denotes a voltage shift due to the trapped charges within the dielectric.

4.1.2 Dielectric Absorption Test (DA)

Measurements of the dielectric absorption loss (as described in detail by Kerber et al.[98]) were carried out as well. The dielectric absorption is one of the effects which is, beside the leakage current, responsible for a charge loss in capacitors[99]. This effect is observable in all dielectric materials [100]. The dielectric

4 CHARACTERIZATION METHODS

absorption tests were performed as follows: the capacitor devices under test are charged for 2 s and discharged for 20 μs (Figure 4.2a). Measured relaxation (Figure 4.2b) is indicating the dielectric losses due to relaxation and trapping. Dielectric absorption values are typically extracted after 64 ms discharge (DRAM refresh time) for different charging voltages [99]. Even though that the mechanism behind the dielectric relaxation is not completely elucidated, the dielectric absorption is modeled with the several capacitor-resistor (RC) elements in parallel to the tested capacitor [99,100]. These RC elements represent the different parasitic effect taking place within the dielectric.

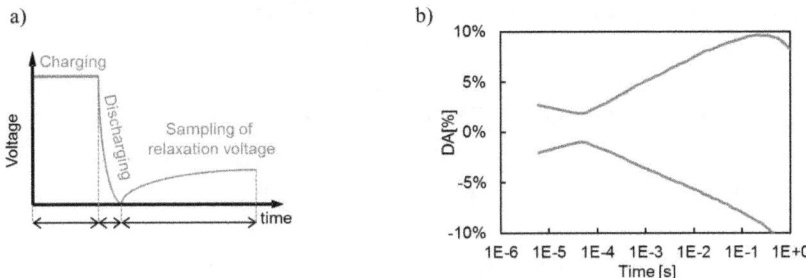

Figure 4.2 a) Charging and discharging voltage behavior during dielectric absorption/relaxation measurements and b) typical dielectric absorption measurement characteristics. DA represents ratio of relaxed and applied voltage.

4.1.3 Breakdown Voltage Test (VBD)

As described in [101] the current voltage characteristics can be measured in a stress and sense scheme with an additional monitor current measurement at a lower voltage. The current voltage characteristic reveals different features dominating at different voltages. This can be seen in the monitor current (I_{sense}) measured at a fixed sense voltage during a voltage sweep (Figure 4.3a). The decrease of I_{sense} at low stress voltage is attributed to electron trapping. The increase of I_{sense} for higher voltages can be explained by a lowering of the tunneling barrier due to electron detrapping (or hole-trapping). For even higher stress voltages finally the hard break down (BD) occurs. Although the current increase at the sense condition is likely dominated by Coulomb barrier lowering, which results in smooth current increase, also a localized leakage path by stress induced leakage current (SILC) or trap assisted tunneling might be present.

4 CHARACTERIZATION METHODS

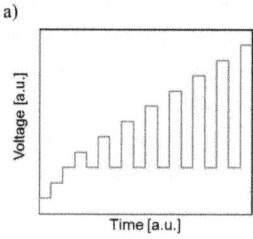 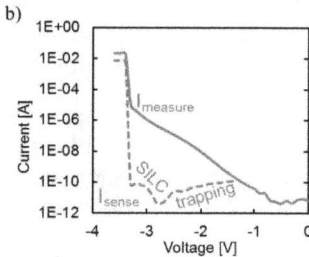

Figure 4.3 a) Voltage transient of the VBD test. VBD test with measured and sense current indicating the trapping and SILC stage.

4.2 Ferroelectric Characterization Methods

4.2.1 Polarization-Voltage Test

The polarization-voltage (P-V) test represents the standard ferroelectric screening method for determination of remnant polarization and the coercive field strength. P-V measurements can be performed on both, capacitor as well as on the transistor structures. Since it is a two point measurements source, drain and bulk terminals are set on the same potential in transistor case. An alternating voltage signal is applied (see Figure 4.4a), while the resulting current is being integrated in order to obtain the polarization response that is finally plotted against the applied voltage as depicted in Figure 4.4b. Details about the circuitry needed for the measurement is based on the Sawyer-Tower circuit described elsewhere[102].

4.2.2 PUND Test

The PUND test is a ferroelectric specific measurement technique used to differentiate the ferroelectric from dielectric response of the current. The method is based on the previously discussed P-V test, but in contrast to the P-V the first voltage pulse if followed by a pulse of the same polarity. The pulse train is completed by the pulse pair with opposite polarity with respect to the first pair (see Figure 4.4c). The first pulse of the pulse train contains the ferroelectric switching current as well as the pure dielectric current component. On the other side, the response on the second pulse contains only a dielectric response due to the fact that the ferroelectric was set in the current state with a previous pulse. By subtracting the current response obtained with the second pulse from the current response obtained with the first pulse, an intrinsic ferroelectric

4 CHARACTERIZATION METHODS

switching current component is obtained. In contrast to the PUND used within the scope of this thesis, a classical PUND comprises trapezoidal pulses.

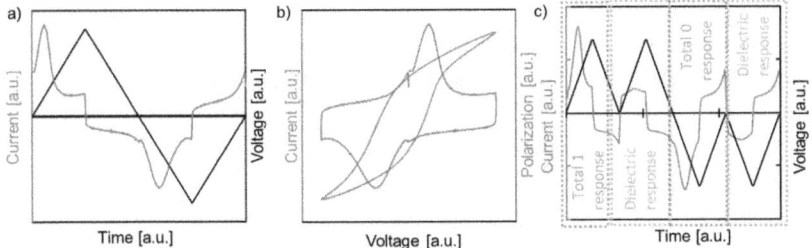

Figure 4.4 Basic ferroelectric characterization methodology. a) Voltage transient for the *P-V* measurement, corresponding current transient response and resulting b) *I-V* and *P-V* characteristics. c) PUND measurement methodology. By subtracting the dielectric response from the total response, intrinsic, leakage-free characteristics of the ferroelectric material is obtained for the positive and negative state, respectively.

4.2.3 First Order Reversal Curve Test

Besides the *P-V* and the *PUND* test used for the screening of the P_r and E_c, the most important test that enables the screening the ferroelectric switching density and internal bias fields is FORC. Even though the method was originally developed to characterize magnetic materials [103,104], it was also successfully applied to typical ferroelectric materials as described elsewhere [105,106,107]. As reported in refs. [22] and [108] the FORC measurement procedure starts at fields high enough to reach the saturation polarization, here the positive one. Subsequently, the field is swept toward the negative direction and back to saturation, stepwise lowering the value of the reversal field until the opposite (here the negative) saturation polarization is reached. This measurement procedure and the derivation of the switching density $p^{\cdot}(E_r,E)$ utilized for probing the reversal fields $E_{r,i}$ in descending order [109] is shown in Figure 4.5a. Resulting frequency distribution in the so called Preisach plane is calculated as the mixed second derivative of the recorded polarization $P^{\cdot}(E_r,E)$ response of the ascending branches of the field sweeps as follows from equation 4.2. The calculation of this switching density as reported by Schenk *et al.* [108] and Pešić *et al.* [22] can be described as follows: For each i-th reversal field $E_{r,i}$ plotted as the y-axis of the Preisach density, the difference in the transient current response to the current response of the previous reversal field $\Delta I(E_{r,i}, E)$ = $I(E_{r,i}, E) - I(E_{r,i-1}, E)$ is plotted versus the field E with $E_r \leq E \leq E_{max}$, i.e. for the field sweep between the reversal field and the maximum field (positive saturation).

4 CHARACTERIZATION METHODS

$$\rho^-(E_r, E) = \frac{1}{2}\frac{\partial^2 \bar{P}_{FORC}(E_r, E)}{\partial E_r \partial E} = \frac{1}{2E}\frac{\partial \bar{j}_{FORC}(E_r, E)}{\partial E_r} \approx \frac{1}{2E}\frac{\bar{j}_{FORC}(E_{r,i}, E) - \bar{j}_{FORC}(E_{r,i-1}, E)}{E_{r,i} - E_{r,i-1}} \quad (4.2)$$

Utilizing mathematical coordinate transform [105,106] given by equation 4.3 switching density as function of the coercive field E_c and the bias field E_{bias} can be derived and plotted.

$$E_c = \frac{E - E_r}{2}, \quad E_{bias} = \frac{E + E_r}{2}, \quad (4.3)$$

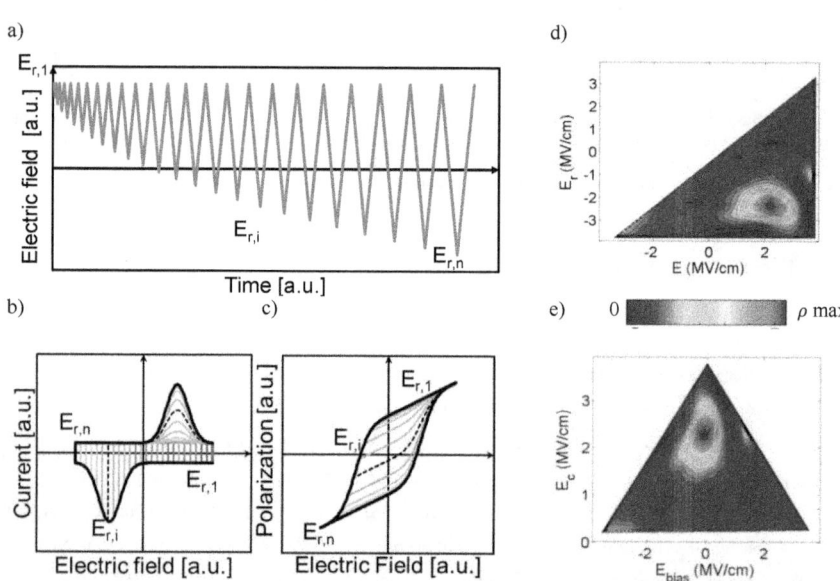

Figure 4.5 FORC approach: a) The excitation electric field transient from positive saturation to obtain $\rho^-(E_r, E)$; b) Resulting current-field and c) polarization-field characteristics. d) calculated experimental Preisach density in E_r-E plain; e) Calculated experimental Preisach density in E_c-E_{bias} plain. For the depicted version of the experimental Preisach density c) using the coordinate transformation given by equation 4.3 it is possible to obtain the Preisach density $\rho^-(E_c, E_{bias})$. Out of the sweeps from the first reversal field $E_{r,1}$ to the last reversal field $E_{r,n}$, one example was chosen as the i-th curve (black dashed lines) to illustrate how the measurement translates to the obtained switching density plot. Figure was reproduced based on study by Pešić et al.[22].

4 CHARACTERIZATION METHODS

4.3 Memory Specific Characterization Methods

4.3.1 Endurance Test

The ability of a memory device to withstand the stress caused by consecutive program and erase operations is characterized by endurance. As a product specification, it defines the maximum number of the PRG/ERS cycles which a certain memory can withstand so that the memory states remain distinguishable. The endurance test is performed as follows: each PRG/ERS operation is followed by the read-out operation which monitors the evolution of the MW with field stress. The typical sequence used for recording of the retention of the characteristic is illustrated in Figure 4.6a. Endurance recorded on the HfO$_2$ based MIM capacitor is shown in Figure 4.6b.

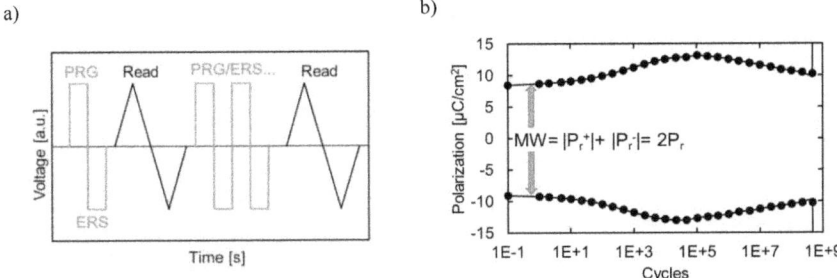

Figure 4.6 a) Sequence of voltage pulses used for recording of the endurance characteristics. b) Endurance characteristics of typical HfO$_2$ based ferroelectric capacitor. PRG and ERS denote program and erase pulse respectively.

4.3.2 Retention Test

In addition to the endurance characteristic, retention of the memory has to be examined. Retention represents the ability of a non-volatile memory to maintain the written state after the removal of the power supply. Retention tests are usually performed at elevated temperatures to provide information about the potential 10-year data retention, which can be extracted by extrapolating measurement values based on suitable models as described in ref [110]. In case of the FeRAM memories, same-state (SS), opposite-state (OS) and new-same state (NSS) retention can be defined [111]. The pulse sequences used for positive and negative state retention tests are illustrated in Figure 4.7a while the calculation of the respective parameters (e.g. OS) is shown in Figure 4.7b. To assess all states by retention tests, four discrete capacitors were used and exposed to different pulse sequences shown in Figure 4.7a. Similar to the PUND tests, the resulting current is dependent on the voltage history and the previous state in which the capacitor resided before the

4 CHARACTERIZATION METHODS

measurement. The first switching pulse comprises dielectric and ferroelectric switching current components, whereas during the second pulse only the dielectric response (including leakage currents and relaxed polarization) is observed. Subtracting the resulting current of the second non-switching pulse from the first one, a normalized switching contribution (or normalized polarization "1" was obtained). The last pulse of each sequence was used to set a defined polarization state, after which the sample was stored at different temperatures for different time intervals. After each bake, the same sequence was applied. As an example, sequences applied to capacitor one and two can be analyzed. There, first pulse of the pulse train, A1 will result in only a dielectric response since the previously written state using A4 is the same. In contrast to capacitor one, the application of the pulse B1 capacitor two yields a current response which comprises ferroelectric switching current and dielectric response since the capacitor resided in the opposite polarization state (last pulse of the sequence two was negative).

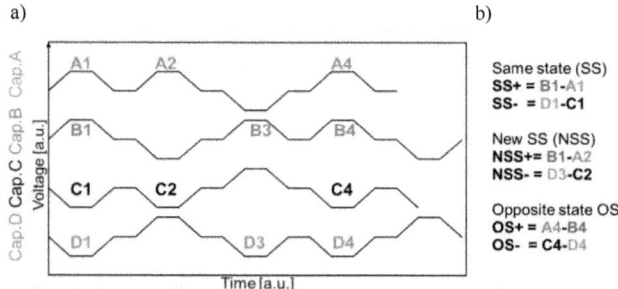

Figure 4.7 a) Sequence of voltage pulses used for same state (SS), new same state (NSS) and opposite state (OS) retention tests. b) Calculation of the retention values specific for FeRAM from the respective integrated current response.

5 Capacitor Stack Properties and their Influence on the Charge Transport

Studying the influence of field cycling on the thin films with spontaneous polarization is complex. A physical interplay governs the purely dielectric mechanisms (i.e defect – O vacancy and O ion – creation, diffusion, recombination, distribution) and ferroelectric properties (change of the remnant polarization by e.g. domain wall pinning, seed inhibition and other modifications of the switching process). Before digging into this strongly entangled system, the intrinsic properties of the pure dielectric response need to be studied. This is necessary to get an idea, on how defects, charged defects and pure dielectric engineering influence the leakage current and further ferroelectric properties within the film. These mechanisms are connected to each other and take place both at interfacial and in bulk regions. With this in mind, the complexity of the observed material system has to be decreased. Thus, the parameter extraction, the consecutive study of leakage mechanisms and the modeling were performed on the thin ZrO_2 based films that do not exhibit a spontaneous polarization. As mentioned in the Chapter 2.1, the binary ("sister") oxides ZrO_2, HfO_2 and other hafnium alloys were considered as promising replacements for SiO_2 among various high-k dielectrics. Due to their large band gap, relatively low leakage current and very good thermal stability [26] their usage became a standard in both capacitors and transistors. For many years, dynamic random-access memory capacitors comprising ZrO_2 based materials have been successfully scaled down to smaller nodes [112,113,114,42]. Metal-insulator-metal capacitors with TiN electrodes and $ZrO_2/Al_2O_3/ZrO_2$ (ZAZ) dielectric stacks have been successfully applied during the last decade in commercial DRAM products from the 80 nm node down to current 18 nm generations [42,43]. Recently, Pešić *et al.* [42,43] reported how the engineering of the existing production type ZAZ layers could be implemented in sub 18 nm nodes. Furthermore, ZrO_2 based films are the currently used in the state-of-the-art DRAM devices that share a similar cell architecture with the FeRAM memories. Moreover, HfO_2 is currently being used as the standard gate dielectric of the CMOS semiconductor industry for 32 nm and below.

Both ZrO_2 and HfO_2 are characterized by very similar values of the dielectric constant, a band gap around 5.7 eV, conduction band offset of at least 1.5 eV to TiN and high defect concentration [26,115,116]. Besides, it is important to note that that neither the introduction of an Al_2O_3 interlayer in case of ZrO_2 based DRAM nor Sr doping of hafnia films does influence the band parameters such as conduction band offset and the band gap [22,43]. Before entering the analysis of the transport within the high-k layers, a detailed material study was needed.

5 CAPACITOR STACK PROPERTIES AND THEIR INFLUENCE ON THE CHARGE TRANSPORT

To address the influence of parameters of the capacitor stack on the charge transport and the device performance the study is organized as following:

1) To rule out the influence of the spontaneous polarization (in Sr doped HfO$_2$ films) and characterize the intrinsic dielectric properties, ZrO$_2$ based DRAM thin films were chosen as a test material.
2) The interface properties, reliability as well as the internal bias fields were studied utilizing TiN/ZrO$_2$ stack with different top electrodes (TiN and Pt).
3) Ahead of the leakage current model development and analysis, a detailed parameter extraction was presented.
4) In the next step, a comprehensive modeling approach framework is developed and introduced to explain the electrical data in terms of physical mechanisms responsible for the leakage current and internal bias field within the ZrO$_2$ based capacitor.
5) Finally, the obtained model was extended to the ferroelectric capacitors based on Sr:HfO$_2$.

Certain parts of the text and results given within this chapter contain the findings presented after the peer-review process [43] and [42].

5.1 Basic Properties of ZrO$_2$ based Metal-Insulator-Metal Capacitors

The leakage current measurements represent, on the one hand, a basic technique, but are, on the other hand, one of the most sensitive approaches for determining the process variations as well as the quality of the dielectric. Leakage currents of TiN/ZAZ with TiN and Pt top electrode are shown in Figure 5.1. The characteristics were recorded sweeping up the voltage in steps of 0.1 V till the breakdown, while the delay of each step was 20 s. This delay time was utilized in order to rule out all relaxation processes and thus, to obtain the intrinsic characteristics of the device. Interestingly, an asymmetric leakage current characteristic is obvious even though top and bottom electrodes are made of the same material - TiN in this case. This unexpected behavior of the, at the first sight, symmetric stack, can be explained by the formation of different interfaces towards the electrodes. Namely, during the deposition of the dielectric a formation of the TiO$_2$ interface takes place between the electrode and the dielectric. During this process TiN pulls oxygen out of the ZrO$_2$ dielectric and nitrogen diffuses into the dielectric [43]. Additionally, the bottom interface layer grows with deposition of the top electrode as reported by Weinreich *et al.* [117]. This interface formation gives rise to the asymmetry of the stack. Therefore, due to the different interfaces (which play a significant role in electron/hole injection in sub 10 nm dielectrics) leakage current measurements result in asymmetric characteristics for positive and negative voltage polarity, i.e. the cases of electron injection from top and bottom electrode, respectively.

5 CAPACITOR STACK PROPERTIES AND THEIR INFLUENCE ON THE CHARGE TRANSPORT

The so-called "capacitance equivalent thickness" (CET) as a basic measurements used for the qualification of the dielectric materials, represents a thickness of the capacitor if it would be built of SiO$_2$. It is calculated according to equation 5.1:

$$CET = k_{SiO_2} \varepsilon_0 \frac{A}{C_{meas}} \quad (5.1)$$

, where k_{SiO_2} represents the permittivity of SiO$_2$, ε_0 the permittivity of vacuum, A the area of the capacitor and C_{meas} the measured capacitance.

In addition to the *I-V* characteristics of TiN/ZAZ stack with different top electrodes, a leakage-CET plot is given in Figures 5.1b. Beside the already observed trends of the current density magnitude for top and bottom injections, a deviation of the CET value can be seen. Even though both stacks were annealed without top electrode, the stack with TiN TE exhibited a lower CET value with respect to the equivalent capped with Pt TE. Beside the oxygen scavenging (formation of the top TiO$_2$ interface) by TiN TE deposition in TiN/ZAZ/TiN MIM on one side, the reason for this fluctuation might partly lay in the poor adhesion between Pt electrode and the ZrO$_2$ dielectric in case of TiN/ZAZ/Pt on the other side. Hence, mechanical contact between the measuring needle could induce stress that modifies the effective area of Pt-capped capacitors, resulting in an inconsistency of the CET calculation. Nonetheless, the influence of adhesion was ruled out by performing statistical measurements.

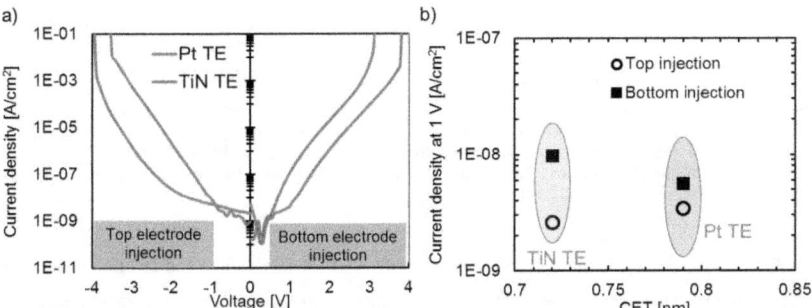

Figure 5.1 a) current voltage *I-V* Measurement of a MIM capacitor (-4 V to +4 V): TiN/ZAZ/ stack with TiN (blue) and Pt (red) top electrode and b) CET vs. leakage current density at 1 V for different top electrode materials.

In order to analyze the influence of the introduction of the noble metal electrode on the stack reliability VBD, dielectric relaxation and the trapping experiments were performed (details in section 4.1). It can be seen in Figure 5.2a that the stress of the device with TiN TE initially caused electron trapping before the onset of stress induced leakage current, finally followed by the hard breakdown of the device. In contrast to the TiN capped stack, the stack with the Pt top electrode exhibited negligible trapping and SILC followed by the hard breakdown. Breakdown voltages were extracted and shown in Figure 5.2b where a clear improvement of the reliability (increased breakdown voltages) was observed for the noble metal electrode with respect to the benchmarking TiN based structure.

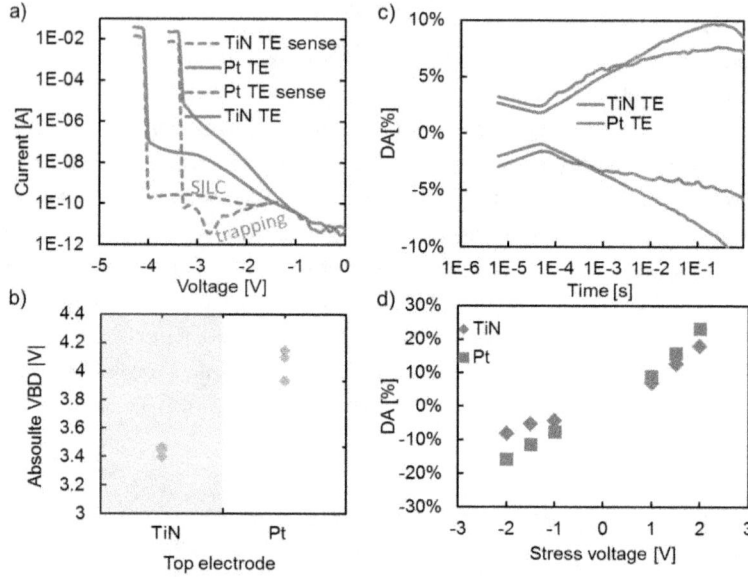

Figure 5.2. A reliability performance comparison of the TiN/ZAZ stack with TiN and Pt top electrodes, respectively. a) I-V characteristic with sensing current (I_{Sense} as described in 4.1.3) measured during a voltage sweep for a 6.5 nm ZAZ/TiN stack annealed at 550 °C with different electrodes on top. b) Break down voltage behavior comparison for different electrodes on ZAZ/TiN stacks. Dielectric absorption c) measured at 1 V and d) extracted at 64 ms for different electrodes at different stress voltages.

The dielectric absorption test was performed for the TiN/ZAZ stack with TiN and Pt TE at different bias voltage conditions. The structures with different top electrodes were stressed at a constant voltage of 1 V and the observed dielectric absorption values are depicted in Figure 5.2c. It can be seen that the sample with top TiN electrode showed the best results of only 6.5±0.3 % DA after 64 ms. Device caped with Pt top electrode exhibited increased DA value and asymmetric characteristics. The workfunction asymmetry most likely also caused the asymmetry in the dielectric absorption graph and increased the value of the dielectric absorption. Although the circuit is shorted, an internal bias field is generated by the different workfunctions of Pt and TiN and disturbs the relaxation currents.

5 CAPACITOR STACK PROPERTIES AND THEIR INFLUENCE ON THE CHARGE TRANSPORT

The internal bias field $E_{built-in}$ generated by using the electrodes with different workfunction values is described by the following equation:

$$E_{built-in} = \frac{1}{d \cdot q} \cdot (WF_{TE} - WF_{BE}) \quad (5.2)$$

, where d represents the thickness of the dielectric, q the elementary charge, WF_{TE} the workfunction value of the top electrode and WF_{BE} the workfunction value of the bottom electrode.

In addition, the voltage dependency of dielectric absorption was extracted and plotted in Figure 5.2d for positive and negative bias condition. The injection side denotes the discrete interface i.e. application of negative voltage on TE causes an injection of electrons from the top electrode, thus represents a spectroscopic result for the top interface. The variability in the values is lower than in the values for bottom electrode injection (positive voltage applied on the top electrode). These results correlate with the leakage current results in Figure 5.1a, which show also larger leakage current differences for different electrodes for positive bias condition. Therefore, it can be concluded that the introduction of the noble metal electrode leads to a decrease of the trapping within the stack, an increase of the breakdown voltage of the stack and an increase of the dielectric absorption component. Generally speaking, this approach improved the robustness of the stack and thus, will be applied for the improvement of the ferroelectric stacks as well (see Chapter 5.3 for details).

Besides reliability, detailed modelling studies of the leakage current mechanisms and the interfaces were performed through the utilization of the different electrodes and parameter extraction described in following subchapters. In the past, the leakage current behavior of these ZrO_2 based MIM capacitors has been studied extensively[49,117,118,119]. Recently it was reported that the dominant leakage current mechanisms of TiN/ZrO_2/TiN (9 nm thick ZrO_2) is trap assisted tunneling [49,120]. The leakage current transport mechanisms are determined by the trap level depth, trap density in ZrO_2 and by Schottky barrier height (SBH). Therefore, detailed characterization and parameter extraction will be presented in the following chapter.

It is important to determine the limits of the intrinsic leakage and to understand the correlation of all the parameters influencing the charge transport. Thus, the MIM structure was modelled using Synopsis TCAD framework as well as a compact modeling approach. In order to develop such models a determination of the leakage governing parameters such as electrode workfunction, energetic and spatial distribution of defect, conduction band offset, and electron mobility is needed. To acquire the determining parameters a detailed electrical characterization was performed. The examined structures were comprised of a ZrO_2/Al_2O_3/ZrO_2 dielectric (free of spontaneous polarization) thickness series sandwiched between two TiN electrodes or a TiN BE and Pt TE (see Chapter 3.1 for processing details).

5.2 Modeling of the Charge Transport within ZrO$_2$ and HfO$_2$ Based High-k Dielectrics

As previously reported in [43,49], electron transport in thin binary oxides is determined by the trap-assisted tunneling current and it is proportional to the amount of the defects between which the electron is being conducted by either hopping or tunneling. Therefore, the determination of the effective tunneling mass is essential for the parameter study and further model development.

5.2.1 Determining Parameters Governing the Charge Transport within ZrO$_2$ and HfO$_2$ Based High-k Dielectrics

Effective Tunneling Mass

For temperatures below 150 K, the leakage current of a capacitor is not temperature dependent [121]. Based on the method of low temperature measurement described by Kaczer *et al.* [121], Knebel *et al.* [122] extracted the effective tunneling mass, m_T in the range of: 0.4-$0.5 \cdot m^*$. These values are comparable to values in literature [121].

Trap Depth

The determination of the trap depth in ZrO$_2$ based dielectrics for different top electrodes was performed based on the procedure described by Knebel *et al.* [123]. Initially, a Poole-Frenkel plot (Figure 5.3 inset) was extracted from temperature dependent leakage current measurements (Figure 5.3a). Accordingly, the Poole-Frenkel plot was used for the trap depth extraction which was determined as 0.7 eV below conduction band. Overall, the change of top electrode (successive deposition of the Pt electrode) has only a minor impact on the trap level since the ZrO$_2$ dielectric was crystallized without the top electrode and top electrode deposition was performed at room temperature. Since the origin of the trap for the Pt electrode on ZrO$_2$ was expected to be similar to the case of TiN electrodes a similar trap depth was assumed. However, the value extracted by this method has to be taken with some precautions. This rather historical method does not include the charge injection into the first defect state from the electrode and could yield unrealistic values of electron hopping frequency and trap depth as well. Therefore, the method presented here is given just for completeness, whereas the exact trap depth and energetic distribution will be extracted backwards from the fitted leakage currents.

5 CAPACITOR STACK PROPERTIES AND THEIR INFLUENCE ON THE CHARGE TRANSPORT

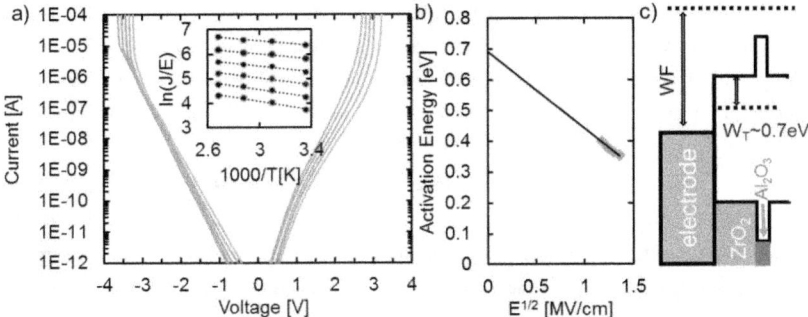

Figure 5.3 Determination of the trap depth for the TiN electrode out of a) temperature dependent leakage current measurements, using (inset) Poole-Frenkel plot. b) The trap depth extracted using Poole-Frenkel plot. c) Corresponding band diagram depicting the extracted energy level.

Trap Concentration

Leakage current as a trap assisted transport is strongly determined by the number of defects within the stack as well as by its spatial distribution. In ref. [43] was reported that the top electrode deposition had significant impact on the trap concentration even though the $TiN/ZrO_2/Al_2O_3/ZrO_2$ stack was crystallized without the top electrode and top electrode deposition was performed at room temperature. In that study, it was reported that the different electrodes caused different properties of the interface as well as a change of the trap concentration within the layer which will be discussed in detail in the modeling section 5.2.1. To address the influence of the top electrode deposition on the defects density, defect spectroscopy was performed as described in the Chapter 4.1. From this measurement, the trap concentration of the sample with the TiN top electrode was determined to be seven times higher than that of the sample with the Pt top electrodes ($7 \cdot 10^{19}$ compared to $1 \cdot 10^{19}$, respectively). This leads to a conclusion that a TiN TE pulls oxygen out of the ZrO_2 dielectric which increases the oxygen vacancy related trap concentration within the stack (Figure 5.4).

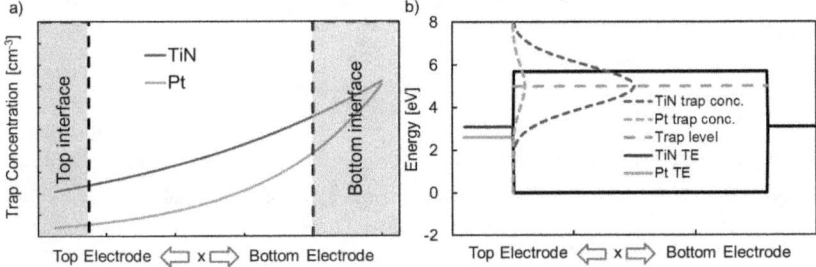

Figure 5.4 a) A sketch of the impact of TiN vs Pt top electrode on trap concentration in the ZrO_2 band diagram. b) Influence of top electrode influence on trap concentration within the capacitor stack [43].

5 CAPACITOR STACK PROPERTIES AND THEIR INFLUENCE ON THE CHARGE TRANSPORT

Analogous to the asymmetric leakage current, the extracted trap concentration varied depending on the polarity of the sweep. Consistent with the higher leakage observed for the bottom injection, a slightly higher defect concentration was extracted. Even though this variation was in the range of the measurement error, it is anticipated that there is a gradient of the defect concentration within the device. This is due to the fact that parasitically grown bottom interface is experiencing more process steps with respect to the top one [117]. Therefore, a vacancy gradient can be illustrated as shown in Figure 5.4a. After the detailed leakage current characterization additional basic parameters of the MIM capacitor based structure, like bandgap, workfunction, and conduction band offset (CBO) between the conduction bands of the metal electrode and the dielectric were determined. To extract these parameters two approaches were utilized: terraced oxide structures were fabricated and characterized at NaMLab gGmbH, while internal photoemission spectroscopy (IPE) measurements were performed at KU Leuven by collaboration partners Nadiia Kolomiets and Valeri Afanas'ev.

Workfunction Extraction Based on Terraced Oxide Structures

Band diagram parameters such as workfunction and conduction band offset determine the injection to the first state by positioning the Fermi level relative to the bandgap. In order to determine the WF value, measurements on terraced oxide [124,125,126] structures are performed. A detailed explanation of sample preparation is given in the Chapter 3.1. The structures prepared by this method are assumed to result in a constant fixed charge value at the interface between the Si substrate and the dielectric. Therefore, the influence of high-k bulk charges is limited to the thickness of ZrO_2-based dielectric. The effective workfunctions (EWFs) for terraced oxide stacks are extracted using equation 5.3, which is a simplified form of the general model given by Wen *et al.* [126] to enable a linear extraction of the V_{FB}–EOT relationship. The contribution of charges in the high-k on V_{FB} is controlled and can be minimized (to ~ +50 meV) by using fixed and thinned high-k film (2-3 nm), thus enabling accurate EWF extraction. V_{FB}–EOT plots for EWF extraction on terraced oxide $Si/SiO_2/ZrO_2/TiN$ is given with Figure 5.5 and equation. 5.3, respectively.

$$V_{FB} = \phi_{ms} - \frac{Q_f}{\varepsilon_0 k_{ox}} EOT \tag{5.3}$$

, where ϕ_{ms} represents the workfunction difference between silicon substrate and metal, Q_f the charge, ε_0 the dielectric permittivity, k_{ox} the dielectric constant of the oxide, and EOT the equivalent oxide thickness of the ZrO_2 high-k material.

The intersect of the *y*-axis on the V_{FB}-EOT dependency shown in Figure 5.5a represents the CBO offset that should be added to the workfunction of the Si (4.05 eV) to calculate the respective workfunction of the top electrode. Finally, the CBO value between electrode and dielectric is calculated by subtracting the electron affinity of the ZAZ film from the corresponding workfunction. Resulting workfunction values of 5.6 and 4.9 eV for Pt and TiN TE, respectively, are well matched with literature values [124]. Finally, the extracted values and corresponding band-diagram illustrations are shown in Figure 5.5 and summarized in Table 5.1.

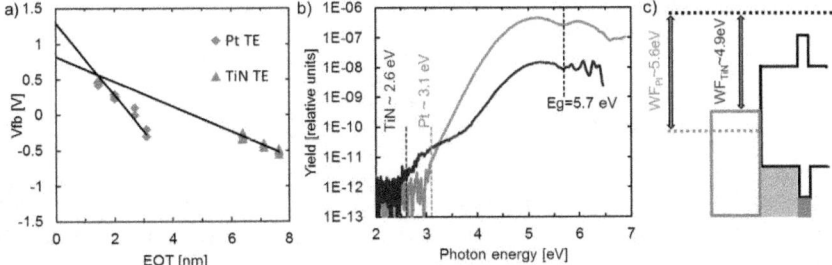

Figure 5.5 a) Flatband voltage (V_{FB}) shift measured on terraced oxide structures vs. the equivalent oxide thickness. b) Photocurrent quantum yield as a function of photon energy as measured on a MIM capacitor stack with indicated positive bias applied to the top metal electrode of a TiN/ZAZ/ stack with TiN and Pt top electrode. c) Corresponding band diagram depicting the extracted energy level. Figure a) taken from ref.[43].

Band Offset and Band Gap Extraction Based on Internal Photoemission Spectroscopy

The internal photoemission spectroscopy (IPE) represents a process of optically induced transition of a charge carrier (electron/hole) from one source (e.g. metal electrode) into another collector (e.g. other metal electrode) across the interface between the two [127]. This method detects every energy offset of the interface as a distinct spectral threshold represented through the minimal energy needed for the injection of a charge carrier. Figure 5.5b shows IPE and photoconductivity (PC) spectra for the TiN/ZAZ/TiN and TiN/ZAZ/Pt capacitors. By observing the photoconductivity spectra, three spectral thresholds can be resolved. The lowest spectral threshold value corresponds to the energy needed for an electron to overcome the CBO between the top electrode and the dielectric. The CBO of the TiN/ZAZ interface is 2.6 eV, which correlates well with the electron affinity of ZrO_2 (2.0 eV [128]) and TiN EWF results of 4.9 eV [124]. The CBO of the Pt/ZAZ interface was determined to be 3.1 eV, which is 0.5 eV higher than the one of TiN/ZAZ. These results show very good correspondence with the previously measured terraced oxide structures, measured in-house [43]. The second spectral threshold should represent the band gap of the interface (IF) layer (TiO_x or TiO_xN_y) formed (by pulling oxygen out of ZrO_2) between dielectric and top electrode and it is measuring ~ 3.4 eV ± 0.2 eV. It should be noted that this spectral threshold is visible only in case of the TiN/ZrO_2 stack capped with TiN TE, whereas the inert nature of Pt electrode suppressed this parasitic. With a further increase of the photon energy a third spectral threshold (corresponding to band gap) is visible.

5 CAPACITOR STACK PROPERTIES AND THEIR INFLUENCE ON THE CHARGE TRANSPORT

Irrespective of the top electrode, the band gap of the dielectric layer is 5.7 eV, which is close to a previously reported value for ALD-deposited ZrO_2 of 5.5 eV [128]. To complete the comparison and justify the approach of usage of ZrO_2-based stack as test vehicle for the parameter extraction, literature values for polycrystalline, ALD-deposited HfO_2 are given in Table 5.1. It can be seen that the extracted values fully correspond with the band diagram parameters of hafnia, reported in [129].

Table 5.1 Band diagram properties of the TiN/ZAZ stacks with TiN and Pt top electrode:

Stack	E_g [eV]	E_g IF [eV]	CBO [eV]
TiN/ ZAZ/IF/TiN	5.7	3.4	2.6
TiN/ ZAZ/Pt	5.7	/	3.1
TiN/ HfO_2/TiN[129]	5.8	/	2.6

Even though the extracted values for the CBO are in the range of the previously reported values, all values seem to be enhanced by 0.1-0.2 eV when a thin interface layer is introduced between the ZAZ stack and the TiN TE.

Accordingly, it can be concluded that the interface layer might plays an important role in the band gap engineering by shifting the Fermi level of the electrode. Therefore, the following mechanism is proposed. The deposition of the TiN top electrodes leads to pullout of the oxygen from the ZrO_2 layer which directly increases the leakage current. The interfacial oxide layer interacts with TiN-based top electrode increasing its workfunction by building an oxide-based, less perfect electrode. The increase of the workfunctions and the consequent shift in the Fermi level enhance the effective barrier height to be surmounted by the electrons. As an alternative, a defined, very thin interfacial oxide could be deposited before the capping of the capacitor with the TiN top electrode. This interfacial layer would serve as an oxygen source which reduces the oxygen defect concentration within the dielectric bulk, increases the barrier to be surmounted by the electrons and consequently decreases the leakage current.

All values determined in this chapter that represent the basic material properties were summed up and shown in Table 5.2. Based on the extracted parameters a band diagram models are derived (see Figure 5.6).

Table 5.2: Overall summary table of relevant parameters for a 6.5 nm thick ZAZ stack deposited on TiN BE and caped with different top electrodes:

TE	CET	WF	CBO	trapping	E_g [eV]	DA^* at -1V	VBD^{**} [V] at 125°C	$J[A/cm^2]$ at 1 V	$J[A/cm^2]$ at -1 V
Pt	0.74	5.6	3.1	weak	5.7	7.5%	-4.1	6.7E-10	2.7E-9
TiN	0.72	4.9	2.6	strong	5.7	4.5%	-3.4	1.8E-8	2.5E-9

Used for simulation in section 5.2   *denotes dielectric absorption **breakdown voltage

5 CAPACITOR STACK PROPERTIES AND THEIR INFLUENCE ON THE CHARGE TRANSPORT

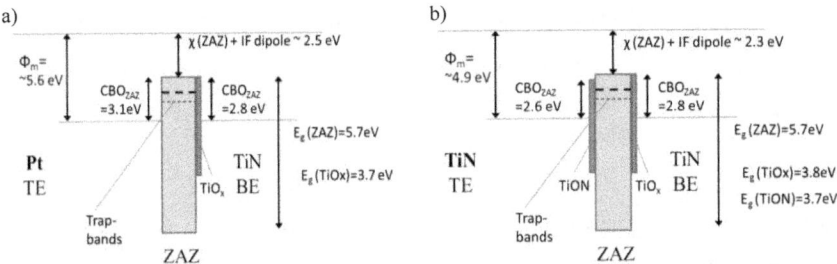

Figure 5.6: Band diagram models extracted from measurements for the TiN/ZAZ stack with a) Pt and b) TiN top electrode. Simplified representation without band banding due to the workfunction difference.

As anticipated from what was discussed in Chapter 5, the introduction of Pt TE results in an improved leakage current behavior compared to the TiN electrodes. Moreover, introduction of noble metal electrode materials showed an improvement in capacitor reliability with respect to the TiN top electrode. Here, the capacitor structure revealed a higher breakdown field. In contrast to other reliability parameters, dielectric absorption tests showed slightly higher values for Pt noble metal electrode (asymmetric stack and internal bias field) with respect to the TiN top electrode.

Furthermore, in regard to the negative bias voltage on the top electrode, the Pt top electrode does not impact the dielectric absorption significantly, whereas the deposition of the TiN TE reduced the dielectric absorption. Comparing the leakage current results, it can be concluded that CBO impacts the leakage currents more, while the dielectric absorption is possibly modified by oxygen vacancies of the bottom layer.

In order to understand the correlation of all the determining parameters it is important to determine the limits of the intrinsic leakage and its influence on device performance. Accordingly, modeling of the MIM structure was performed. Modeling was performed using both a compact modeling approach and Synopsis TCAD.

5.2.2 Charge Transport Modeling Using a Compact Modeling Approach

As previously discussed, high-k materials contain a significant number of defects such as dislocation and impurities which act as electron traps. In order to take those trap states into account they are modeled as single level acceptor and donor traps within the ZrO_2. Those levels are uncharged when unoccupied while they carry the charge of one electron when they are fully occupied. Inside the physics section of the Sentaurus Device command file, trap distributions and trapping models were specified. Traps were defined within the bulk of the material, as well as at the interface between two materials or two regions. Trap

5 CAPACITOR STACK PROPERTIES AND THEIR INFLUENCE ON THE CHARGE TRANSPORT

occupation dynamics is determined by the trap occupation function $f^{n/p}$ which for electron and hole traps can take values between 0 and 1. Change of occupancy $\frac{\partial f^n}{\partial t}$ due to the emission or capturing of electrons[1] can be represented as a sum of all emission and capturing processes r_i^n:

$$\frac{\partial f^n}{\partial t} = \sum_i r_i^n = \sum_i (1 - f^n) c_i^n - f^n e_i^n \qquad (5.4)$$

, where i denotes the number of capture and emission processes, c_i^n represents the electron capturing rate and e_i^n represents the electron emission rate.

Emission and capturing rates implemented were based on nonlocal tunneling. Within the Sentaurus Device nonlocal tunneling to traps is modeled as the sum of an inelastic, phonon-assisted and an elastic process [130,131]. In this study Nasyrov's multiphonon mediated trap-assisted-tunneling model [53,132,133] described in Chapter 2.2 was utilized to model the charge transport effects inside the dielectrics.

Full 3D TCAD-based simulations of the charge transport and consequent calibration is a rather time-consuming process. Hence, before starting the TCAD model development, a charge transport model was implemented in MATLAB to narrow the parameter set for calibration of the TCAD model and to assess its behavior.

Compact Charge Transport Model of $ZrO_2/Al_2O_3/ZrO_2$ MIM Capacitor

Temperature-dependent leakage currents measured on the ZAZ-based MIM capacitors were used as a reference in the first stage of the model development. The compact model was developed using analytic expressions described in Section 2.2. Total current density J_{tot} is composed of Schottky emission J_{SE} (equation 2.1), direct tunneling J_{DT} (equation 2.2), Fowler-Nordheim tunneling (equation 2.3) the modified (tunneling mediated) Poole-Frenkel transport (equations 2.4 and 2.5) and trap-assisted tunneling J_{TAT} (equation 2.7) current components and was modeled with following relation:

$$J_{tot} = J_{SE} + J_{DT} + J_{FN} + J_{TAPF} + J_{TAT} \qquad (5.5)$$

The focus of this section and modeling approach is analysis of the conduction band engineering and its influence on the charge transport. Even though introduction of the interlayer materials as Al_2O_3 blocks the propagation of grain boundaries along the whole stack and consequently reduces the leakage current, it has a minor impact on the CBO value and the trapping properties [134]. Therefore, the charge transport within this simple model will be treated as transport through the single layer, while the impact of Al_2O_3 interlayer

---

[1] Equivalent equations can be written for hole traps, where n is exchanged by p.

5 CAPACITOR STACK PROPERTIES AND THEIR INFLUENCE ON THE CHARGE TRANSPORT

is accounted for through the field adjustments[2] within the stack. The developed compact model was set-up by using the parameters from Table 5.2 and was fitted to the temperature dependent leakage currents. The calibration was carried out for a 6.5 nm thin, ZAZ stack sandwiched with TiN top and bottom electrode (see Section 3.1 for details). The comparison of the measured and simulated leakage current density is shown in Figure 5.7.

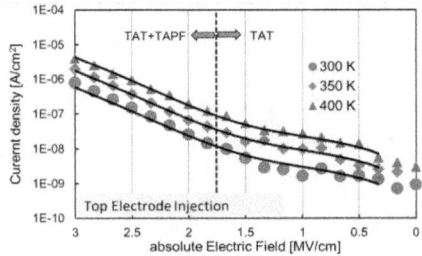

Figure 5.7 Comparison of measurement data (symbols) and simulation results (lines) for MIM capacitor structure with 6.5 nm thick $ZrO_2/Al_2O_3/ZrO_2$ dielectric and TiN electrodes. TAT and TAPF denote the dominating transport mechanisms.

The simulation was performed for temperatures of 300 K, 350 K and 400 K. By positioning the trap band at 1.25 eV (thermal ionization energy) below the conduction band the TAT transport mechanism (equation 2.7) showed a full fit within the range of 0.5 to 1.8 MV/cm (see Figure 5.7). In addition, the trap-assisted Poole-Frenkel (TAPF) transport mechanism, modeled by combining equation 2.4 and equation 2.6, together with TAT current component showed a very good fit within the 1.8 to 4.2 MV/cm field range. This TAPF contribution was described by adding a second defect band at around 0.7 eV which became dominant current component (electrically active)[3] for fields higher than 1.8 MV/cm. A high defect concentration of $5 \cdot 10^{19}$ cm^{-3} was utilized for the simulation.

In the next step, the developed model is applied to the same TiN/ZAZ (6.5 nm thick) stack but capped with Pt top electrode (see Figure 5.8). The TiN/ZAZ capacitor with Pt top electrode showed a similar behavior, only with the observed current components shifted to higher fields. This shift partially comes from a decrease of the trap concentration[4] which consequently yields sparser distribution of defects[5]. Moreover, it was observed that fields below 3.5 MV/cm (or 2 MV/cm in the TiN-case) cannot induce a field high enough

---

[2] Voltage divider due to the different k-values of stack constituents, top ZrO_2, Al_2O_3 interlayer and bottom ZrO_2.
[3] Charge transport through the defect level located at 0.7 eV below conduction band represented a significant current component compared to the contribution of the trap band located at 1.25 eV below conduction band.
[4] In contrast to the TiN TE, inert noble metal electrode is not pulling the oxygen from the dielectric.
[5] Due to the decrease of the defect concentration higher electric field is required for injection of electron.

to significantly contribute to the trap-assisted Poole-Frenkel transport current component and therefore can be assigned to a trap assisted tunneling based leakage. As mentioned above, modeling of low field region was performed using Nasyrov's multi-phonon trap-to-trap transport mechanism [53][132][133] with deep traps at 1.25 eV (2.5 eV optical ionization) below the conduction band of ZrO_2 [43]. The increased workfunction of Pt compared to TiN decreases the probability of electrons tunneling to a trap level due to the increased energetic distance between the Fermie-level of the electrode and the trap level. The decrease of the trap concentration together with the internal bias field (due to the workfunction difference) leads to the observed field-wise shift of the TAPF current component in the case of Pt electrodes. The impact of the increase of the Schottky barrier is also visible as a shift of the Fowler-Nordheim tunneling current component to higher fields, which becomes dominant at 4.2 MV/cm and 5.2 MV/cm for the cases of TiN and Pt top electrode, respectively.

Table 5.3: Simulation parameters for a TiN/6.5 nm ZAZ/TiN capacitor stack. *denotes fitted parameters.

TE	WF [eV]	CBO [eV]	W_t [eV]	E_g [eV]	Trap conc. [cm^{-3}]
Pt	5.6	3.1	0.7/1.25*	5.7	$5·10^{18*}$
TiN	4.9	2.6	0.7/1.25*	5.7	$5·10^{19*}$

It can be seen that the decrease of the trap concentration from $5·10^{19}$ to $5·10^{18}$ cm^{-3} together with an internal bias field due to the asymmetric WF values leads to a leakage current decrease by about two orders of magnitude. A similar behavior is seen for an increasing energetic position of the trap band with respect to the conduction band of ZrO_2.

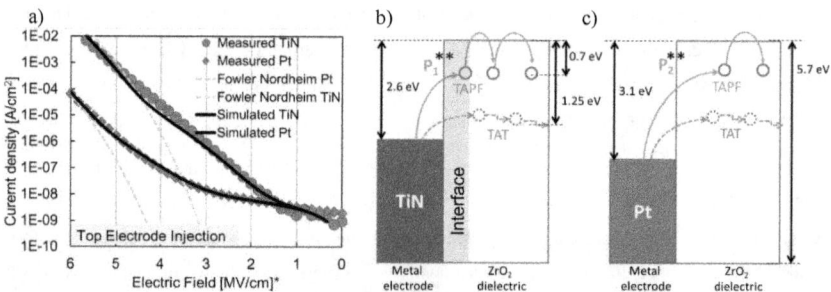

Figure 5.8 a) Modeled leakage current (solid line) and the I-V measurement (symbols) of a MIM capacitor (0 V to -4 V): TiN/ZAZ/ stack with TiN and Pt Top electrode. Suggested model of a band structure of b) TiN/ZAZ/TiN and c) TiN/ZAZ/Pt capacitor (Al_2O_3 band gap and TiN BE were omitted) plotted together with suggested transport mechanisms. *The field was plotted with absolute value. **probability P_1 determining injection from the TiN electrode is higher than probability P_2 denoting the injection from Pt top electrode. Figure taken from study published by Pešić et al. [43].

5 CAPACITOR STACK PROPERTIES AND THEIR INFLUENCE ON THE CHARGE TRANSPORT

After fulfilling the limit of a minimum CBO derived from the Dushman equation (as calculated by Jegert et al.[49]), every further increase of the CBO reduces the Schottky emission current component and its contribution to the total current. Simulations of the CBO increase of 0.5 eV showed a decrease in Schottky emission current by about eight orders of magnitude. However, the amplitude of the simulated Schottky current component was too small and therefore its influence can be neglected. The suggested sketch of charge transport in TiN/ZrO$_2$ stacks with TiN and Pt TE illustrating the previous results and discussion is given in Figure 5.8b and Figure 5.8c.

From the leakage current simulations it can be seen that with an increase of the workfunction and the according CBO the energetic distance between the Fermi level and the trap level increases. This leads to a decrease of the tunneling probability and the TAPF current component. In addition, leakage current simulations revealed that an increasing Schottky barrier height results in a decrease of the FN tunneling leakage at high electric field. Therefore, the Schottky barrier height improvements by the introduction of an inert metal electrode with higher CBO represents one of the most effective ways to reduce the leakage current in the high-k materials and scale down ZrO$_2$-based dielectrics for future capacitor based memories.

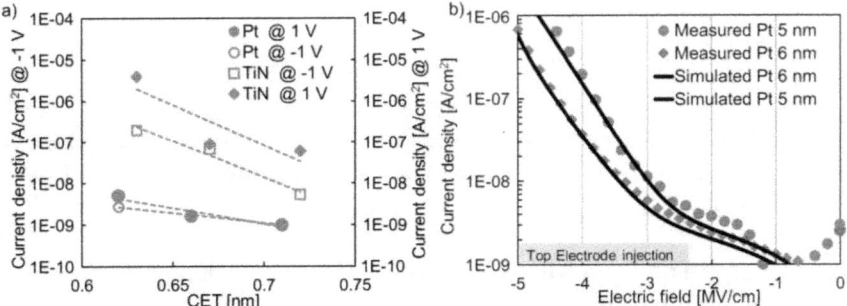

Figure 5.9 a) CET vs. leakage current density at ±1 V for Pt and TiN as top electrode materials. b) Modeled leakage current (solid line) and the I-V measurement (symbols) of a MIM capacitors with TiN/ZAZ/Pt stacks for two different thicknesses of ZAZ 5 nm and 6 nm. Figure taken from from Pešić et al. [42].

To confirm the validity of the developed model and the charge transport mechanisms suggested to govern the leakage through the dielectric, devices with dielectric thicknesses of 5 nm, 5.5 nm and 6 nm with TiN bottom and a Pt as well as a TiN top electrode were examined and characterized. The introduction of the Pt electrode showed improved leakage currents and a decrease of the leakage current for all sample thicknesses. The most important fact is that the leakage current criterion of $1 \cdot 10^{-7}$ A/cm^2 at 1 V was fulfilled even for the thinnest (5 nm) ZAZ layers. From Figure 5.9a it can be seen that samples with TiN top electrodes meet the leakage specification up to a CET value of 0.68 nm, while the use of a Pt top electrode resulted in a reduction of leakage current by almost two orders of magnitude and the target leakage criteria are fulfilled at least down to CET values of 0.62 nm. From the trend, it is anticipated that even thinner films could be used if Pt electrodes are employed [42]. Again, the introduction of the high-workfunction Pt top electrode caused a field shift towards higher fields and a decrease of the leakage for thicker films. In order to validate the model, a simulation of the 5 nm thick ZAZ stack was performed and compared with the simulation of a 6 nm thick dielectric (see Figure 5.9b). Due to the same annealing condition for both thicknesses, the resulting level of crystallization differs. Thinner films are more difficult to crystalize due to the increased crystallization temperature. [123] Therefore, a thinner dielectric shows a lower crystallinity and a lower k-value [42,123]. As a result of the reduced number of grain boundaries, the higher-k material possesses a lower number of defects that behave as electron traps and contribute to the conduction. Moreover, due to the different ratio of the interlayer to the overall ZAZ stack thickness, a slightly lower optical dielectric constant was assumed. Using the parameters listed in Table 5.4, a good model was obtained that corresponded well with the measurement. Similar to the 6.5 nm thick ZAZ capacitors, a trap assisted component (field < 2.5MV/cm) was followed by TAPF (for fields > 5 MV/cm).

Table 5.4: Simulation parameters as extracted from a TiN/ZAZ/Pt capacitor stack. *denotes fitted parameters.

Pt TE	WF [eV]	CBO [eV]	W_t [eV]	E_g [eV]	ε_{opt}	Trap conc. [cm^{-3}]
ZAZ 6 nm	5.6	3.1	0.75/1.3*	5.7	5*	5·10^{18}*
ZAZ 5 nm	5.6	3.1	0.75/1.3*	5.7	4.85*	1·10^{18}*

As discussed before, the reaction of the Pt top electrode with the dielectric is minimal and can be neglected. Here, the concentration of the oxygen vacancies stays constant whereas the workfunction value is increased. This results in lower leakage current for both voltage polarities.

5.2.3 Charge Transport Modeling Using a TCAD Modeling Approach

In order to prepare the model for the analysis of the charge trapping-ferroelectric switching interplay, the developed compact model was transferred to the model implemented within TCAD Sentaurus Device.

TCAD Charge Transport Model of ZrO$_2$/Al$_2$O$_3$ /ZrO$_2$ MIM Capacitor

After narrowing the parameter selection and calibration, the previously developed compact charge transport model of ZAZ-based MIM should be validated. Subsequently, the validated model will be extended to ferroelectric HfO$_2$ and prepared for simulations and analysis of dielectric-ferroelectric interplay behavior in Chapter 6. According to the TEM micrograph (see Figure 5.10d), the ZrO$_2$-based capacitor model, shown in Figure 5.10a-c, was defined using the Synopsys TCAD Structure Editor [54].

5 CAPACITOR STACK PROPERTIES AND THEIR INFLUENCE ON THE CHARGE TRANSPORT

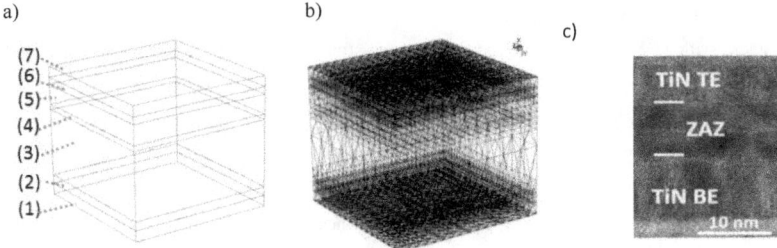

Figure 5.10 3D geometry of the MIM capacitor device. a) MIM geometry consisting of the bottom (1) TiN electrode, bottom interface layer (2), bottom ZrO_2 layer (3), Al_2O_3 interlayer (4), top ZrO_2 layer (5), top interface layer (6) and variable (TiN or Pt) top electrode (7). b) TCAD device mesh with densely meshed regions at interfaces. c) TEM micrograph of 10 nm thick ZAZ layer sandwiched between two TiN electrodes.

The leakage current is modeled including nonlocal tunneling in and out of traps within the dielectric. The nonlocal tunneling model is available within the Sentaurus Device of Synopsis TCAD and unifies direct, Fowler-Nordheim, trap assisted tunneling mechanism[6] and multiphonon emission [53]. The leakage current is calculated using a nonlocal tunneling model set-up in TCAD and taking into account the main material parameters (dielectric constant, barrier and band offsets, hole and electron effective tunneling masses, and the effective dielectric thickness). Furthermore, the TCAD model includes additional parameters such as trap depth, trap density, trap cross-section, capture and emission times.

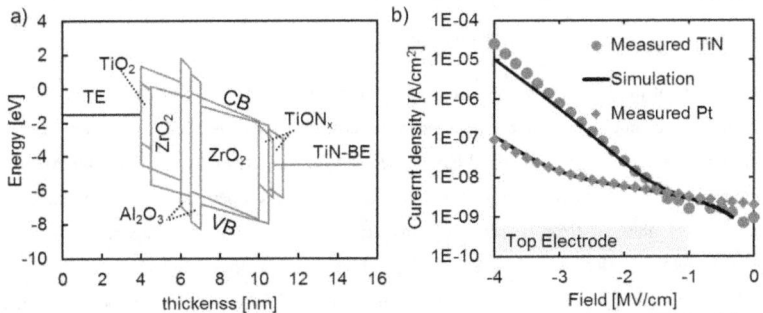

Figure 5.11 Simulation and modeling results. a) TiN/ZAZ/ stack with TiN (blue) and Pt (red) TE band diagrams at -4V applied at the top-electrode defined within the Sentaurus device. b) Simulated I-V characteristics of the TiN/ZAZ stack with TiN and Pt top electrode. Valence and conduction band are denoted with VB and CB, respectively.

---

[6] TAT mechanism implemented within the nonlocal tunneling of TCAD is the very same TAT model of Nasyrov, described in section 2.2.2.

5 CAPACITOR STACK PROPERTIES AND THEIR INFLUENCE ON THE CHARGE TRANSPORT

Using the same parameter set extracted from the compact modeling approach, I-V characteristics of TiN/ZAZ/ stacks with TiN and Pt top electrode were successfully reproduced (Figure 5.11b). The TAPF mechanisms developed in this work, is not available in Sentaurus Device. Thus, a pure trap-assisted tunneling mechanism was utilized to fit the characteristics. This resulted in a shift of the shallower trap band to 1 eV below CB (previously: 0.7 eV). Accordingly, optical ionization energy utilized for the fit was 2 eV below CB (previously: 1.4 eV). In contrast to the compact model, a discrete, non-stoichiometric (parasitically grown) low-k TiO$_x$ interface layers (with defect concentration of 9e^{19}cm^{-3}) were added[7]. These sub-stochiometric regions possess a high value of defects [117]. Consistent with the interface formation reported by Weinreich *et al.*[117] and as discussed before, a top interface layer was thinner than its bottom counterpart. This parasitically grown layers are anticipated to play a significant role in the analysis of the charge-ferroelectric interaction and will be discussed in Chapter 6.

It can be concluded that the determining parameters of the intrinsic (spontaneous polarization-free) dielectric properties were investigated. The performed simulations confirmed the strong influence of CBO on the leakage current and reliability of the stack. With the increase of the workfunction and consequently CBO, the energetic distance between the Fermi level and the trap level increases. This WF increase leads to a decrease of the tunneling probability and dielectric leakage. Therefore, conduction band offset improvements by introduction of inert metal electrodes with higher CBO represent one of the most effective ways that can be applied to improve the leakage and the reliability of the capacitor based memories. The application of this approach and its implications on ferroelectric properties will be discussed in section 5.3.

TCAD Charge Transport Model of Sr:HfO$_2$ MIM Capacitor

After the analysis of the intrinsic dielectric properties, charge transport and interface effects of the ZrO$_2$ dielectric stack which does not exhibit a spontaneous polarization, the developed charge transport model is used to simulate the ferroelectric, Sr doped HfO$_2$ based capacitors. Due to the fact that HfO$_2$ and ZrO$_2$ are often called sister oxides (they possess very similar band diagram properties), [31,32] the band diagram properties can be inherited from the charge transport model of ZrO$_2$-based capacitor. Therefore, all parameters for the WF, CBO and band gap are imported from the previous detailed characterization of the non-switching ZrO$_2$ films. In the Structure Editor of the Synopsis TCAD 3D device structure and mesh are defined analog to the ZrO$_2$ based model (see Figures 5.12a-c). In the next step, distinct bottom and top regions of TiON$_x$ and TiO$_x$ are defined (see Figure 5.12a). Hence, the defined structure comprises a TiN/TiO$_x$/Sr:HfO$_2$/TiON$_x$/TiN stack. In addition, a multiple grains are defined within the Sr:HfO$_2$ layer to address the polycrystalline morphology of the films and its influence on the ferroelectric switching (Figure

---

[7] Interface effect in the simple compact model is accounted via the distance of the first trap from the electrode.

5.12c). Depending on the crystallinity level, the *k*-value and consequently the field drop across the distinct grain can strongly vary and thus, influence the total ferroelectric switching behavior of the film. A detailed discussion is given in the Chapter 2.3 alongside explanations of the implementation of the Preisach based model of hysteresis within Sentaurus Device. Similar to the previously developed model for charge transport in ZrO_2 based thin films, the leakage current is the modeled including nonlocal tunneling in and out of traps within the dielectric. In addition to the previously acquired parameters, the trap spectroscopy, as described in Chapter 4.1, was performed to determine the initial defect concentration. From the trap spectroscopy measurements, a defect concentration of around $3.8 \cdot 10^{19}$ cm^{-3} was determined.

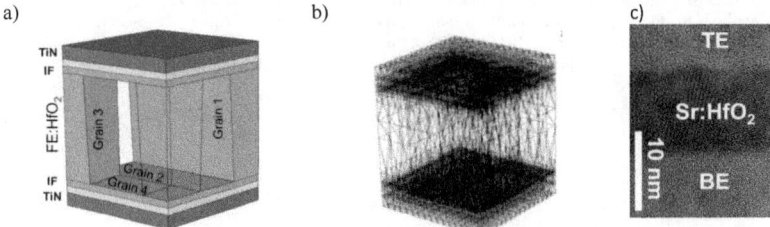

Figure 5.12 3D geometry of the MIM capacitor device. a) The geometry of the developed 3D grain boundary model of the FeCap structure consisting of the bottom TiN electrode, bottom interface layer, Sr:HfO$_2$ layer, top interface layer and TiN top electrode. b) The TCAD device mesh with densely meshed regions at interfaces. c) The TEM micrograph of 10 nm thick Sr:HfO$_2$ layer sandwiched between two TiN electrodes.

The defined geometry and the developed model were used to simulate *I-V* characteristics of Sr:HfO$_2$-FeCaps measured in the same manner as the ZrO$_2$ based counterparts (step size of 0.1 V; delay 20 seconds). Utilizing band diagram parameters extracted from the ZrO$_2$ based stack together with the defect properties and parameters listed in Table 5.5, *I-V* characteristics of the Sr:HfO$_2$ based MIM were successfully reproduced. In contrast to the measured value, an initial density of O vacancy defects of $5 \cdot 10^{19}$ cm^{-3} ($9 \cdot 10^{19}$ cm^{-3} within the interface) was extracted from the calibrated model. Defect concentrations extracted from the temperature dependent leakage current simulations are, as expected, slightly higher than the one extracted with trap spectroscopy. This is due to the fact that the latter extraction method is limited since it considers only the defects below a certain injection energy that are being filled at the chosen sensing voltage.

Table 5.5: Band diagram parameters used in simulation and simulation parameters as extracted from a TiN/HfO$_2$/TiN capacitor stack. *denotes parameters extracted from the simulation.

TiN/Sr:HfO$_2$/TiN	WF [eV]	CBO [eV]	W_t [eV]	E_g [eV]	Trap conc. [cm^{-3}]
10 nm	4.8	2.6	1.2/1.7*	5.7	$5 \cdot 10^{19}$*

5 CAPACITOR STACK PROPERTIES AND THEIR INFLUENCE ON THE CHARGE TRANSPORT

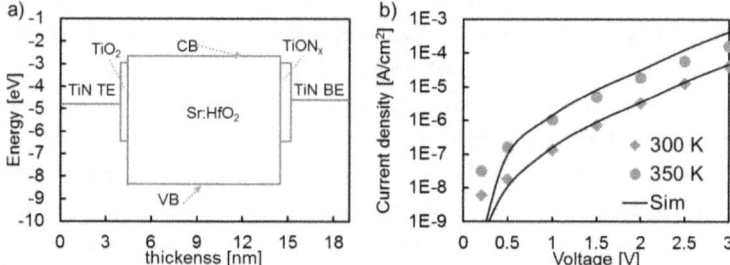

Figure 5.13 TCAD Simulation and modeling results of Sr:HfO$_2$ based FeCap. a) TiN/Sr:HfO$_2$/TiN band diagram at +0.2 V applied to the top-electrode with a single trap band defined within the Sentaurus device. b) Simulated I-V characteristics of the Sr:HfO$_2$ based FeCap with TiN top and bottom electrodes. Valence and conduction band are denoted with VB and CB, respectively

5.2.4 Charge Transport Model of Sr:HfO$_2$ MIM Capacitor Developed by MDLab Software

For the purpose of the analysis of the field cycling behavior of the FeCap, simulation of the oxygen vacancy movement (discussed in detail in Chapter 6.2) was performed at the University of Modena and Reggio Emillia (UniMoRE). During the one month long visit to Prof. Larcher's group at the UniMoRE, an equivalent (to TCAD developed) compact charge transport model of Sr:HfO$_2$ based FeCap was implemented and calibrated. This model was utilized for the simulation of vacancy motion as will be explained later in Chapter 6.2. In the following a calibration of the charge transport model developed within the MDLab [135] software package is given.

Figure 5.14a shows the model geometry of TiN/Sr:HfO$_2$/TiN based FeCap, comprising O vacancies (red circles) implemented in the commercially available MDLab tool. The model was calibrated by performing the fit (see Figure 5.14c) of the temperature-dependent leakage currents [21] measured on the pristine FeCap using the multiphonon trap-assisted-tunneling (TAT) transport model introduced in Chapter 2.2 and readily available in the MDLab package [51]. Given that the charge transport is determined by the defects within the layer and that the grain boundaries and defects are randomly generated within the layer during the formation of the capacitor device, their spatial and energetical distribution is non-uniform. It is important to note that in contrast to the TCAD[8] model that allows only discrete energetic levels, in the MDLab simulation package a defect band with a random spatial and energetic distribution can be generated (Figure 5.14b). An initial density of O vacancy defects of $5.4 \cdot 10^{19}$ cm^{-3} ($9 \cdot 10^{19}$ cm^{-3} within the interface) with a defect ionization energy distribute in a defect band spreading from 1 eV to 3.2 eV below the

---

[8] Newest distribution (April 2016) of Synopsys TCAD allows definition of discrete energetic and spatial coordinates for defects.

5 CAPACITOR STACK PROPERTIES AND THEIR INFLUENCE ON THE CHARGE TRANSPORT

conduction band was extracted. These values were used for the subsequent modeling and investigation described in Chapter 6. They correlate with complex transport mechanism based on dominating trap-assisted-tunneling within the HfO_2 dielectric that was reported elsewhere [55,56,129,136]. It should be noted that, besides the vacancies, that actively assist the electron TAT current[55,56], interstitial oxygen ions were considered within the stack. Moreover, in contrast to the model implemented in the TCAD, here each defect are fitted according the algorithm reported in [51] so that each has its own, unique capture/emission time and relaxation energy.

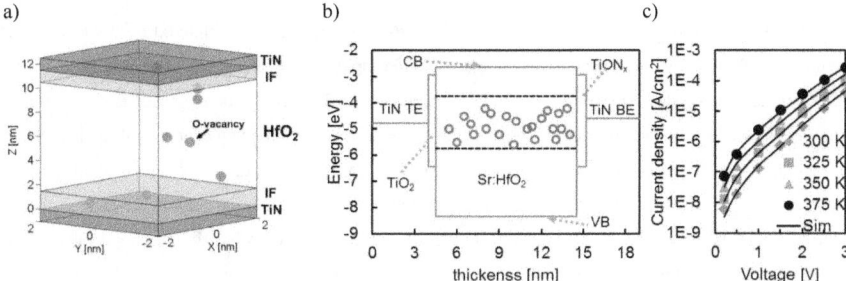

Figure 5.14 MDLab software based simulation and modeling results of Sr:HfO$_2$ based FeCap. a) Model geometry of TiN/Sr:HfO$_2$/TiN based FeCap with oxygen vacancies (big red spheres). b) FeCap band diagram simulated at 0 V. Circles denote defect states (oxygen vacancies) distributed within the boundaries of trap band (dashed lines). Valence and conduction band are denoted with VB and CB, respectively. c) Calibrated simulation (solid line) of the temperature dependent I-V characteristics (symbols) of the Sr:HfO$_2$ based FeCap. Figures a) and c) are taken from the study performed by Pešić et al. [22].

The calibrated charge transport models of Sr:HfO$_2$-based MIM capacitor, implemented in TCAD (described in sections 5.2.3) and implemented in MDLab software (section 5.2.4) will be utilized in Chapter 6 to track the defect concentration evolution with field-cycling of the ferroelectric capacitor and its influence on the ferroelectric properties of the film. The model implemented in MDLab software will be broadened (see Chapter 6.2 for details) and employed to investigate the leakage current, O vacancy generation, diffusion and its evolution with field cycling. The extracted defect densities[9] will be transferred to extended TCAD[10] model of ferroelectric capacitor to investigate the ferroelectric properties of the stack. Before starting the comprehensive modeling study, applicability of the techniques that led to enhancement of the DRAM stack properties such as interlayer introduction as well as the usage of the inert noble metal electrode should be investigated. Therefore, the following subsection is devoted to the influence of the noble metal electrode introduction and consequent stack asymmetry on dielectric and ferroelectric properties of the device.

---

[9] By fitting the model to the stress-dependent leakage current, defect density will be extracted.
[10] TCAD model of Sr:HfO$_2$ MIM capacitor developed in section 5.2.3 will be extended by adding ferroelectric switching properties based on Preisach model described in section 2.3.2.

5.3 Introduction of Internal Bias Fields

As learned from the ZrO$_2$ based stacks (section 5.1), the introduction of the high WF noble metal electrode like Pt is beneficial since it resulted in: a reduction of the leakage current and consequently, an increase of breakdown voltage and reliability of the film, as well as in a decrease of oxygen scavenging from the stack and thus, less formation of interfacial layers between dielectric and the electrodes. In following, benefit of Pt top electrode introduction and its influence on the dielectric and ferroelectric characteristics will be addressed.

After the crystallization of TiN/Sr:HfO$_2$/TiN stacks, TiN top electrode was removed by a wet chemical etching step. The removal of TiN TE was followed by a Pt top electrode deposition. In Figure 5.15a an *I-V* characteristic comparison between the Sr:HfO$_2$/TiN stacks with TiN and Pt top electrode is given. Analog to the ZrO$_2$ based stack, the introduction of the high WF electrode resulted in a decrease of the leakage for both voltage polarities. Besides the influence on the interface properties and consequently, an asymmetric charge transport through the stack, this electrode exchange strongly influences the ferroelectric properties of the film. The asymmetry introduced by the electrodes with different workfunctions changes the potential landscape of the Helmholtz free energy. As a result, asymmetric switching and backswitching conditions are expected. Using equation 2.11 (discussed in detail in Chapter 2.3), a symmetric and an asymmetric stack were simulated (Figure 5.15b). Asymmetric electrodes create a different height of the barrier that has to be overcome for polarization switching and backswitching to positive polarized and negative polarized state, respectively. Similar to the field shift observed in the static *I-V* measurements, the dynamic *I-V* characteristics and consequently the *P-V* characteristics of the TiN/Sr:HfO$_2$ stack with Pt top electrode became asymmetric. A shift of the switching peaks and both coercive fields to more positive values with respect to the stack equivalent with the TiN top electrode was observed (Figure 5.15c).

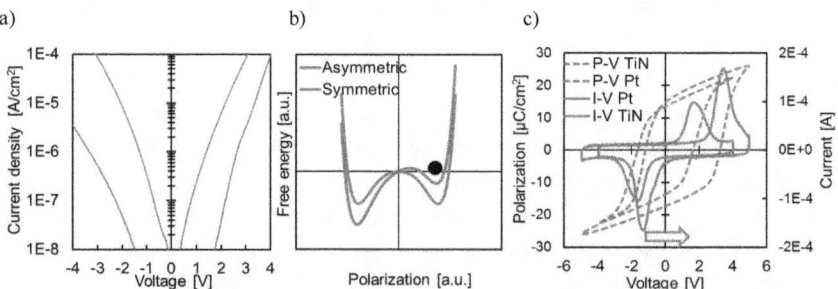

Figure 5.15 Comparison of the asymmetric (Pt TE) stack and symmetric (TiN TE) stack properties. a) Static leakage current for the TiN/Sr:HfO$_2$ stack with TiN and Pt top electrode. b) Sketch of free energy of the TiN/Sr:HfO$_2$/ stack with TiN (blue) and Pt (red) top electrode and corresponding c) *I-V* (solid) and *P-V* (dashed*)* characteristics.

According to equation 2.11, asymmetric electrodes, i.e. top electrode characterized with the higher workfunction, would effectively increase the barrier for switching to positive side, whereas they decrease the barrier for backswitching to the negative side. The internal bias field of the different electrodes favors the negative over the positive polarization state. Hence, it is expected that switching current peaks and the corresponding P-V hysteresis undergo a field shift towards more positive fields. Figure 5.15 evidences that the asymmetric electrodes introduced an internal bias field which resulted in a decrease of the leakage current and shift of the switching current peaks towards more positive fields. However, it can be seen that instead of a pure field-wise shift, the positive E_c is significantly more influenced than the negative E_c. This means that the hysteresis becomes broader (E_c+ - E_c- increased), which indicates that also the switching kinetics get modified. Müller et al. (for capacitors) [137] and later Mulaosmanovic et al. (for transistors) [138] reported that switching within polycrystalline Si doped HfO_2 films is governed by the nucleation limited switching (NLS) kinetics. From the NLS switching kinetics [139][140][141] it can be seen that nucleation of the domain is enhanced with increasing electric field. Therefore, a internal bias field generated by the WF difference and stack asymmetry would impede the domain nucleation for one polarity, but enhance it for the opposite polarity. This assumption could be verified by carrying out access time measurements [142] in a future work, where a reference pulse is used for setting the positive state and the pulse of the opposite polarity is parametrized (pulse amplitude and length is varied) in order to study its impact on the switching process to the opposite, i.e. negative, state. Further, measurement should be repeated in such manner so that parameterized pulse and reference pulse possess inverse polarities with respect to the first run. In case that differences in access time for switching and backswitching could be observed this would be a strong indication that internal bias field, deepening on its direction impede or facilitate the nucleation kinetics.

5.4 Summary

In this chapter, mechanisms responsible for the charge transport within the high-k films were discussed. Initially, in order to decrease the complexity of the ferroelectric-dielectric interplay a ZrO_2 based MIM capacitors were chosen for the parameter extraction since:

a) ZrO_2 and HfO_2 are considered to be twin oxides characterized with similar band diagram parameters [26],

b) utilization of ZrO_2 based MIM enables addressing intrinsic dielectric nature and ruling out the ferroelectric switching parasitics,

5 CAPACITOR STACK PROPERTIES AND THEIR INFLUENCE ON THE CHARGE TRANSPORT

c) techniques that led to enhancement of the DRAM stack properties such as interlayer change [134] as well as the usage of the inert noble metal electrode [43,42] could be applied for improvement of more complex ferroelectric stacks in both 1T-1C and 1T configurations.

Initially, to address interface properties, their influence on the charge transport and reliability, different top electrode materials X (Pt, TiN) were compared in a TiN/ZrO$_2$/Al$_2$O$_3$/ZrO$_2$/X MIM capacitor structure. Pt led to improved leakage current behavior compared to TiN electrodes. This noble metal electrode material showed an improvement in capacitor reliability with respect to the TiN top electrode. With this approach, the capacitor structure revealed a higher breakdown field. On the other hand, dielectric absorption tests showed slightly higher values for noble metal electrode with respect to the TiN top electrode, which can be explained by the stack asymmetry that introduces an internal bias field. These techniques enable the enhancement of the spontaneous polarization-free stacks and can be applied to improve, from the governing mechanism point of view complex, ferroelectrics stack. Beforehand, based on the extensive characterization and parameter extraction, the charge transport of the MIM capacitor stack was modeled. To narrow the parameter set needed for calibration of a complete 3D TCAD model, a compact model was developed. Further, extracted band diagram parameters were used for the derivation of the 3D TCAD charge transport model of the ZrO$_2$-based MIM to address intrinsic, spontaneous polarization-free dielectric.

Finally, the carefully evaluated charge transport model was extended and applied to the ferroelectric, Sr doped HfO$_2$ system. The model was implemented in TCAD in order to address the interplay of the trapping and ferroelectric switching process. In parallel to the TCAD modeling the temperature dependent leakage currents were fitted with the commercially available MDLab dielectric simulator. Further, guided by the same approach like in the ZAZ case, introduction of the noble metal electrode decreased the leakage currents, thus increasing: a) reliability from the dielectric point of view and b) number of domains that could be reached by increasing the breakdown strength of the films and enabling usage of the higher fields. In the following chapter, the extended model will be used to simulate the oxygen vacancy drift/diffusion processes due to field cycling and their influence on the switching characteristics of the ferroelectric.

6   Field Cycling of the Ferroelectric High-*k* Materials

After the detailed analysis of the charge transport and determining parameters responsible for the reliability of the stack, an extension of the previously developed model is needed to address central reliability issues of ferroelectric-based memories. Field cycling stability represents one of the major requirements of any memory. To compete with sophisticated, state-of-the-art memory stacks as the ZAZ, shown in Chapter 5, each cell of the new, emerging solution has to withstand a stress caused by program and erase operations for target specific 10^{12} cycles[11]. Therefore, this chapter is focused on a detailed analysis of the field cycling influence on both dielectric and ferroelectric properties of the ferroelectric stack. Even though PZT-based ferroelectric memories are known for the long retentions and easily fulfill the 10-year target, the stability of the ferroelectric properties of hafnia-based ferroelectric with field cycling is insufficient [21,22,143]. The memory window, defined as the difference between positive and negative remnant polarization[12] for a capacitor-based cell, usually shows a pronounced evolution: After an initial increase, which itself is already undesired, a strong degradation starts quite early with respect to the specified 10^{15} cycles. Within the endurance characteristics of the ferroelectric capacitor, two distinct lifetime stages can be isolated (see Figures 2.16b and 6.1b):

    a) "Wake-up": The increase of the remnant polarization P_r corresponds to an opening or de-pinching of the pristine pinched hysteresis loop with field cycling [144,145];

    b) "Fatigue" or aging: After a certain number of cycles, the remnant polarization starts decreasing, which results in a closure of the memory window due to fatigue mechanisms [146].

Wake-up and fatigue mechanism occur due to a complex interplay of phenomena involving both ferroelectric (e.g. domain orientation, phase stability and field induced phase transitions) and dielectric properties (e.g. defect activation/creation, charge injection and consequent trapping), that could induce internal bias fields [108]. Physical mechanisms behind wake-up and fatigue of conventional ferroelectric materials were studied in detail during the past 50 years. A huge number of publications proposed numerous scenarios dealing with increase and decrease of the polarization in perovskite-based materials. Different studies of wake-up on PZT material system were performed and reported previously [144,147,148]. Furthermore, Tagantsev *et al.* in [149,150] and Lou *et al.* in 2009 [146] presented very detailed and systematic reviews of fatigue mechanisms. These studies suggested that a combination of several mechanisms (domain de-/pinning [151], seed inhibition [152], and formation of the passive/dead

---

[11] Embeded Flash (eFlash) requires 10^5 cycles [192][182].
[12] The remnant polarization is a figure of merit of 1T-1C cells. Depending on the state of the cell, remnant polarization defines the difference in charge flowing to the bit line between the two memory states 1 and 0.

6 FIELD CYCLING OF THE FERROELECTRIC HIGH-K MATERIALS

layer[153]) could be responsible for the device wake-up and fatigue. Even though those research efforts give important hints at important points to look for also in new materials, they were confined to potential mechanisms separately and in a speculative manner. Therefore, in this Chapter, based on [21,22] a detailed study of the physical mechanisms responsible for the wake-up and fatigue stage in doped hafnium films is presented out. A comprehensive collection of experimental results derived from the extensive electrical characterization of the wake-up and fatigue phenomena will be presented in Section 6.1 and used for development of a physical model of ferroelectric capacitor. Within the Sections 6.1-6.3 characterization results will be interpreted, for the first time, using a novel modeling approach, which will allow to point to the root cause of wake-up and fatigue phenomena in 10 nm thick strontium doped hafnia based ferroelectric thin films. To address these fundamental questions, a detailed internal bias field screening evolution with field cycling is presented in Section 6.2. This is a necessary initial step to get an idea of how charged defects and ferroelectric properties are distributed within the film. Moreover, a sketch of the potential interplay of defect mechanisms to be aware of is derived from characterization results and compared to literature. With this in mind, Section 6.3 is devoted to investigating the physical mechanisms behind the wake-up and the fatigue phenomena. Furthermore, a comprehensive modeling framework is developed and introduced to explain the electrical data in terms of physical mechanisms responsible for the field cycling effects within the ferroelectric capacitor.

Certain parts of the text within this chapter contain the findings presented after the peer-review process in [21], [22] and [154].

6.1 Wake-up Behavior in the Ferroelectric Memories

As it can be seen in Figure 6.1a, pristine ferroelectric devices based on doped HfO_2 are usually characterized by double peaks in the current voltage characteristic and consequently, a pinched hysteresis loop (see Figure 6.1b). Schenk *et al.* [94] reported that pristine double peak behavior may originate from different internal bias fields within the device stack [94]. It is assumed that different internal biases or a different field manifestation on certain grains and even portions of the stack may occur due to a non-uniform defect distribution, polycrystallinity of the annealed film and *k*-value non-uniformity [154]. To address these fundamental questions and investigate the mechanisms behind, a detailed characterization and assessment of the mechanisms of the wake-up and fatigued stages will be presented. Characterization was carried out on the MIM capacitor structures with ALD-deposited 10 nm thick $Sr:HfO_2$ dielectric sandwiched between the two TiN electrodes and crystalized in N_2 atmosphere at 800 °C for 20 s (for details please see the Section 3.1 and Figure 3.2c).

6 FIELD CYCLING OF THE FERROELECTRIC HIGH-K MATERIALS

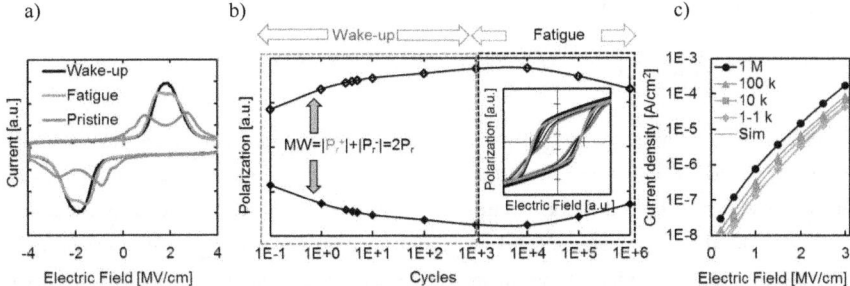

Figure 6.1. a) Current-voltage and corresponding (b-inset) polarization voltage characteristics of the ferroelectric Sr:HfO$_2$ for pristine, woken-up and fatigued state. b) Extracted evolution of positive and negative remnant polarization with field cycling. The difference between positive and negative remnant polarization, P_r^+/P_r^-, is related to the memory window. c) Leakage current evolution measured after different numbers of cycles with lines depicting the simulation (Sim). Figures taken from studies by Pešić et al. [21, 22].

I-V curves are measured on the ferroelectric capacitor at three stages of lifetime: a) pristine; b) woken-up; c) fatigued stage (Figure 6.1a). It can be seen that the transient *I-V* characteristic measured on the pristine device is characterized by two distinct current peaks, which move towards each other with continuous bipolar cycling, eventually merging during the wake-up regime. Due to the fact that polarization-voltage characteristic represents integral of the transient current, this behavior acts as a transition from the initially pinched hysteresis loop to a open hysteresis (Figure 6.1b-inset) [155]. Further cycling results in a broadening of the current peak and consequent decrease of the P_r amplitude in the fatigue stage.

In order to get insight into intrinsic dielectric properties of the stack, the evolution of static DC leakage current with bipolar cycling was monitored in parallel to the endurance behavior (Figure 6.1c). In addition to monitoring of the static leakage, leakage current defect spectroscopy was employed. As mentioned before, leakage current defect spectroscopy represents a powerful technique for examining the quality of a dielectric in terms of defects generated/activated within the dielectric layer during the bipolar cycling stress. Surprisingly, magnitude of leakage does not exhibit any noticeable increase up to 10^3 cycles (the wake-up stage). However, further cycling resulted in a significant increase of the leakage current as shown in Figure 6.1c. This increase indicates a generation of new, electrically active, defects within the dielectric stack during cycling. In contrast, the constant leakage current characteristics of the wake-up phase pose that there is almost no generation of new defects affecting the charge transport governed by TAT. However, there might be a redistribution of the existing ones that gives rise to the changes in *P-E* characteristics. In addition to a defect redistribution, a change of its occupancy could take place as well. To explore these possibilities

6 FIELD CYCLING OF THE FERROELECTRIC HIGH-K MATERIALS

FORC was employed, which is sensitive to changes in internal bias fields that might be generated by changes in defect distribution or occupancy.

6.1.1 FORC and Internal Bias Screening

Just recently, Schenk and co-workers [108] utilized FORC as a very powerful tool to analyze the evolution of the switching and back-switching field distributions in hafnia-based ferroelectrics. Even though it is used for the characterization of ferroelectrics for decades [105,106,107], the FORC approach originates from the research of the ferromagnetic materials [103]. A detailed [105,106] and simplified mathematical treatment of the measured data, a coordinate transform to E_c and E_{bias} and measurement procedure were described in Chapter 4.2.3.

Within the scope of this thesis, the FORC technique is applied to analyze the evolution of the local internal bias field and coercive field evolution with field cycling of the device. As with the *I-V* characterization, the focus of the analysis will be on the discussion of certain pristine, woken up and fatigue stages (see Figure 6.1a). Utilizing the coordinate transformation (see Chapter 4.2.3) the obtained Preisach densities ρ for these three stages are plotted as a function of bias field E_{bias} and coercive field E_c in Figure 6.2.

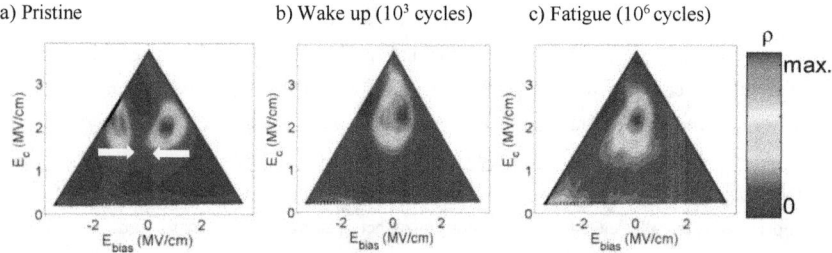

Figure 6.2. Evolution of the Preisach/switching density ρ during device cycling after: a) 1; b) 10^3 and; c) 10^6 cycles corresponding to the pristine, wake-up and fatigue stage, respectively. Figures taken from studies by Pešić et al. [22]

As shown in the Preisach/switching density plot (Figure 6.2a), the pristine sample is characterized by two distinct switching density populations, with two opposite bias fields and almost same coercive field. The observed behavior fully corresponds to the results of the dynamic hysteresis measurements depicted in the inset of Figure 6.1a. Similar to results observed by Schenk et al. [108], the local internal bias of these regions disappears with field cycling resulting in movement of the Preisach/switching density maxima toward each other. As depicted in Figure 6.2b the maxima of the switching density finally end up merged around zero bias, following the same trend seen for the current peaks in the *I-V* characteristics in Figure

6 FIELD CYCLING OF THE FERROELECTRIC HIGH-K MATERIALS

6.1a. The FORC approach was able to prove the initial existence of the local internal bias fields and their disappearance with further cycling.

The increase of the transient current peaks and the consequent increase of the magnitude of the merging peaks manifests in an increase in the remnant polarization observed in dynamic hysteresis measurements. This P_r increase indicates an increase of the number of domains participating in the switching process, whereas the disappearance of the local internal bias field suggests that the field distribution within the stack tends to become more uniform with field cycling. Moreover, the change of the internal bias could be related to the distribution changes of defects within the stack. These defects can either act as: a) trapping centers that change the internal field consequently by changing their occupancy states (e.g. oxygen vacancies [21]) or b) local changes of the material either by phase transformation or vacancy movement under the influence of field and temperature.

Further cycling results initially in the onset of the fatigue behavior, followed by progressing decrease in P_r. The internal bias screening and analysis during the fatigue stage, was carried out utilizing measurement procedures in a similar manner as for the woken-up case. After the device was stressed with alternating switching pulses for 10^6 cycles FORC measurements were recorded. From the pseudo-color plots in Figures 6.2b-c (wake-up and fatigue case) it can be seen that the size of Preisach density decreases. Moreover, this decrease is accompanied by slight biasing of the sample towards negative fields, pointing on an increase of asymmetry in the TiN/Sr:HfO$_2$/TiN device stack. In general, it is expected that the dielectric degradation (which primarily corresponds to the O vacancy generation [156,157] and consequent charge trapping result in a decrease of the effective electric field seen by the ferroelectric layer. This, consequently, reduces the number of switchable domains resulting in a decreasing memory window.

As previously discussed, the gradual decrease of the local internal bias field can be attributed to the wake-up effect. In this discussion, it was suggested that a redistribution of O vacancy defects and a change of their charge state can occur. Due to the fact that vacancy mobility is governed by the electric field and temperature, the evolution of the P_r was recorded at high temperatures. Namely, the increase of the temperature should decrease the energy barrier needed to be overcome by a vacancy in order to move through the lattice during the wake-up stage. In Figures 6.3a-b, it can be seen that with increasing temperature the number of the cycles needed to wake-up the sample decreases significantly. This strong temperature dependence of the wake-up indicates that diffusion/drift process as well as the defect generation can be responsible mechanisms [22]. As mentioned before, the leakage current, however, stays constant (see Figure 6.1c). This hints at the fact that the generation of new defects is not the dominating process even though it cannot be completely ruled out.

6 FIELD CYCLING OF THE FERROELECTRIC HIGH-K MATERIALS

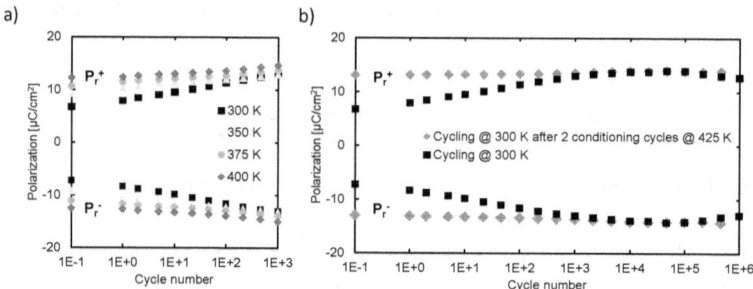

Figure 6.3. a) Thermal wake-up dependence on elevated temperatures. b) Thermal wake-up conditioning procedure to be used in novel FeCap devices.

Encouraged by the strong temperature influence on the wake-up of the ferroelectric capacitor, conditioning experiments were performed. The strong temperature acceleration of the wake-up phase suggests to utilize of higher temperatures to precondition the ferroelectric memory devices to achieve stable and uniform device performances. Hence, FeCap devices were subjected to just two cycles at elevated temperatures after which the endurance test, i.e. stress and $P-E$ measurement at room temperature were performed. Following the previously discussed trend, the number of the wake-up cycles reduces significantly and completely disappears at 150 °C preconditioning temperature as shown in Figure 6.3b. Later on, this approach was used by Fengler *et al.* [143] to extract activation energies for wake-up. The extracted values were in the range of 0.5 eV[13], nicely corresponding to the activation energy needed for moving of the O ion [158]. However, those values from literature span a rather wide range of at least half an order of magnitude. [94,148,159]. Thus, further substantiation is needed.

6.1.2 Structural Changes and Transmission Electron Microscopy Study

To investigate field cycling influence on the ferroelectric layer morphology a TEM study was conducted. This study was conducted by a collaborating partner from the North Carolina State University at group of Prof. James LeBeau. TEM lamellas were prepared using a focused ion beam (FEI Quanta) with final thinning at 2 kV 30 nm thick Gd:HfO$_2$ MFM stacks were prepared for STEM. STEM study was conducted with a probe-corrected FEI Titan G2 60-300 kV operated at 200 kV with a beam current of approximately 80 pA, a probe semi-convergence angle of approximately 19.6 mrad, and a HAADF detector with inner semi-angle of approximately 77 mrad[22]. The structure of the Gd:HfO$_2$ is expected to be similar to the

---

[13] Activation energy determination based on this procedure might be influenced by the leakage current, therefore a tracing of the E_c instead of P_r is recommended. Just recently, Fengler *et al.*[193] reported an activation energy for diffusion of 1 eV which is identical to the activation energy characteristic to PZT-based films [148].

electrically tested Sr:HfO$_2$, but its thickness is higher, which for this first extensive TEM study seemed a preferable choice. These Gd doped HfO$_2$ films show analog electrical behavior to the Sr doped HfO$_2$ samples investigated in this work as already reported by Hoffmann *et al.* [160]. In order to remove drift distortion, images were acquired using the RevSTEM technique. [161] The acquisition involved averaging of approximately 40 frames sized 1024 x 1024 pixels with 90° rotation between frames and a 2 μsec dwell time. Before frame averaging, images with excessive distortion were removed manually after drift correction [22]. Lamellas for STEM examination were prepared by a focused ion beam from MFM capacitors with 30 nm thick Gd:HfO$_2$ in pristine, wake-up, and fatigue stage. Utilizing STEM partial phase transformation and structural changes were observed. Figure 6.4 shows STEM images of the HfO$_2$/electrode interfaces for each cycling stage. As previously suggested, a co-existence of multiple phases, monoclinic (paraelectric, PE), tetragonal (PE) and orthorhombic (ferroelectric, FE), was observed. During wake-up, the monoclinic phase portion within the film decreased by transforming into orthorhombic phase. In addition, the observed tetragonal HfO$_2$ regions adjacent to the electrodes are less present in the cycled samples as well (see Figure 6.4b-d). It can be seen that the device in the pristine stage exhibits the greatest tendency to relax towards tetragonal symmetry [22]. Also, the wake-up image shows some relaxation towards the interface. However, it seems to retain the phase of the bulk grain more than the pristine. In contrast to the pristine and wake-up stage, fatigued sample shows little to no relaxation at the interface. In addition to the influence on the ferroelectric switching the observed phase changes have also significant impact on the dielectric constant of the complete stack and the field distribution within the stack. Cubic phase of HfO$_2$ exhibits *k*-values of between 30 and 39 [69,33,162], tetragonal between 28 and 70 [33,69,34], orthorhombic (Pca2$_1$) has *k*-value in range from 27 to 35 [69], while monoclinic phase is characterized with *k*-values between 16 and 20. Since the field scales inversely with *k*-value local changes between e.g. monoclinic and orthorhombic can easily result in about 50 % less field drop at this location of the stack.

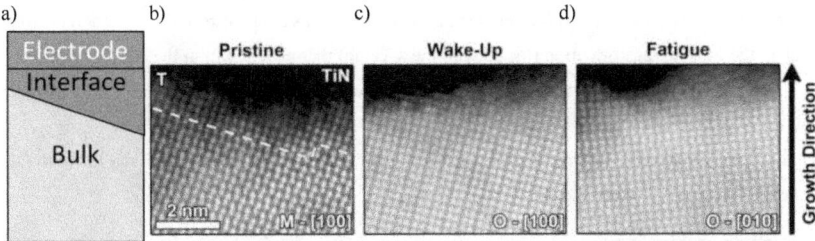

Figure 6.4 Single-grain segments of the capacitor stack near the electrode interface. a) Sketch of the interfacial region next to the electrode. HAADF-STEM images of doped HfO$_2$ in b) pristine, c) wake-up, and d) fatigue conditions showing relaxation of the bulk monoclinic (PE) and tetragonal (PE) symmetry at the electrode interfaces towards orthorhombic (FE) phase. This trend which is most pronounced in the pristine and least pronounced in the fatigued. Intensity levels are adjusted to enhance contrast near the electrode. Figures b-d taken from study by Pešić *et al.*[22].

6 FIELD CYCLING OF THE FERROELECTRIC HIGH-K MATERIALS

To address this important question, small-signal capacitance-voltage measurements (±3 V carrier DC signal with 30 mV AC modulation at 100 kHz) were performed. Similar to the previous experiments, the capacitance evolution was recorded during field cycling. Standard ferroelectric C-V characteristics exhibit a butterfly-like hysteresis shape (Figure 6.5). The maxima of the butterfly hysteresis occur due to the contribution of domain wall capacitance around the coercive field (during the polarization switching) [163]. Since the contribution of the domain wall capacitance in the saturated region of the characteristics is minimal, the extraction of the effective dielectric constant of the complete stack including interfaces and bulk was performed in these regions[14]. With cycling, a continuous drop of the total capacitance/k-value by 4 % was observed (see Figure 6.5).

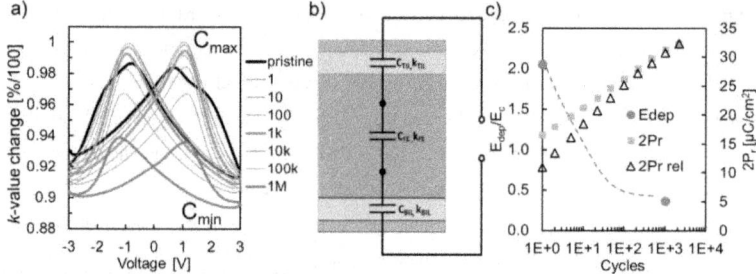

Figure 6.5 a) Evolution of the butterfly-shaped curve of small-signal capacitance with field cycling. The peak value of the capacitance C_{max} and minimal value C_{min} are marked. The former gives an indication of the amount of charge inside the system and the latter is used to monitor the k-value evolution, which might give a hint at phase changes in the film stack. b) Electrical model of the MFM stack: Equivalent circuit consisting of three series capacitances. c) Ratio of depolarization and coercive field (E_{dep}/E_c) in comparison to normalized P_r and relaxed P_{r_rel} after one second for doped HfO$_2$ sample. Figures a)-b) taken from study by Pešić et al. [22] and c) reproduced from [164].

Comparing the capacitance evolution with the TEM results, this behavior indicates that a T→O phase transformation (decrease of the k-value) in the interfacial region has a much stronger impact on the total k-value than the M→O transformation (increase of the k-value) that occurs within the bulk. A simple (first order approximation) electrical model consisting of three serial capacitors representing two interfaces C_{IF} and bulk C_{FE} of the stack is given in following.

By applying Gauss' law to the interface of the three dielectric slabs the following equation for the three capacitors in series can be expressed as:

---

[14] It should be noted that due to the breakdown strength of the capacitor dielectric as well as due to the high stress induced by the C-V measurement, a complete saturation of the C-V is not possible, and thus slight influence of the domain wall capacitance might be sensed. On the other side, the wake-up effect and corresponding domain de-pinning would yield a higher capacitance values, as seen for voltages than 2 V. Nonetheless, this was not the case on the maximum voltage (Voltage range >2 V) and therefore, the influence of the domain wall capacitance at maximum fields can be neglected in further analysis.

6 FIELD CYCLING OF THE FERROELECTRIC HIGH-K MATERIALS

$$\frac{1}{C_{eq}} = \frac{1}{C_{BIL}} + \frac{1}{C_{FE}} + \frac{1}{C_{TIL}}\bigg|_{C_{BIL}=C_{TIL}=C_{IF}} \quad (6.1)$$

$$\frac{1}{C_{eq}} = \frac{2}{C_{IF}} + \frac{1}{C_{FE}} = \frac{2C_{FE}+C_{IF}}{C_{IF}C_{FE}} \quad (6.2)$$

$$C_{eq} = \frac{C_{IF}C_{FE}}{2C_{FE}+C_{IF}}\bigg|_{C=k\varepsilon_0\frac{A}{d}} \quad (6.3)$$

$$\frac{k_{eq}}{d_{eq}} \sim \frac{\frac{k_{IF} \cdot k_{FE}}{d_{IF} \cdot d_{FE}}}{2\frac{k_{FE}}{d_{FE}} + \frac{k_{IF}}{d_{IF}}} \quad (6.4)$$

$$k_{eq} \sim \frac{\frac{k_{IF} \cdot k_{FE}}{d_{IF} \cdot d_{FE}}}{2\frac{k_{FE}}{d_{FE}} + \frac{k_{IF}}{d_{IF}}}, \quad (6.5)$$

, with C_{eq} denoting the equivalent capacitance, C_{BIL} denoting the capacitance of the bottom interface layer, C_{FE} denoting the capacitance of the ferroelectric layer, C_{TIL} denoting the capacitance of the top interface layer, A, d, and k denote the area, thickness and dielectric constant of the capacitors, respectively. Consequently, d_{eq}, d_{IF}, d_{FE} and k_{eq}, k_{IF} and k_{FE} denote the thicknesses and dielectric constants, respectively of the equivalent, interface and ferroelectric bulk capacitor.

The whole stack can be represented as a voltage divider of three capacitors in series as depicted within the Figure 6.5b. The final equation 6.5 represents a first order approximation of the total dielectric constant, describing a system consisting of three homogeneous capacitors in series. Here, a direct influence of the interface and ferroelectric layer on the total k-value and corresponding capacitance can be concluded. For completeness, different phases, their polarization properties and corresponding dielectric constants are given in Table 6.1. With the transformation of the tetragonal (paraelectric) towards orthorhombic (ferroelectric) phase, the k-value is reduced. Besides this fact, the transformed portion of the films becomes ferroelectric and generate an additional contribution to the total ferroelectric switching current. Consequently, the electric field over this device fraction compared to other lateral positions rises, which enables the material to reach a higher remnant polarization value. In the pristine stage of the device both monoclinic fraction of the bulk and tetragonal interfaces are present. As a consequence, a higher amount of the applied voltage drops over the monoclinic portions of the stack as they are characterized by a lower dielectric constant. Therefore, the rest of the ferroelectric bulk experiences reduced effective electric field, thus resulting in lower polarization values.

Table 6.1 Overview of the different phases within the polycrystalline HfO$_2$ thin films and its properties [34][69][162].

Bulk Phase	k-value	Properties	Field (across bulk)
Monoclinic (M-phase)	~18	Paraelectric	Higher than for O-phase
Orthorhombic (O)	~30	Ferroelectric	
Tetragonal (T)	~35	Paraelectric	Lower than O-phase

6 FIELD CYCLING OF THE FERROELECTRIC HIGH-K MATERIALS

Another parameter that can be traced and possibly correlated with structural changes is the depolarization field. To address the influence of field cycling on the depolarization field, an endurance characteristic was recorded. In addition to monitoring the remnant polarization P_r, the relaxed remnant polarization P_{r_rel} (P_r after one second of waiting time) was recorded. To address the decrease of the paraelectric phase fraction with field cycling a depolarization field [165][71][166] is calculated[15] (equation 6.6).

$$E_{dep} = \frac{P_r}{k_{FE}\varepsilon_0}\left(1 + \frac{C_{IF}}{C_{FE}}\right)^{-1} = \frac{P_r}{k_{FE}\varepsilon_0}\left(1 + \frac{k_{IF}d_{FE}}{k_{FE}d_{IF}}\right)^{-1} \qquad (6.6)$$

, where E_{dep} represents the depolarization field, P_r the remnant polarization, C_{IF} the interface capacitance; C_{FE} the capacitance of FE layer; d_{IF} the interfacial paraelectric layer thickness and d_{FE} the FE layer thickness as measured by TEM (dielectric constants: $k_{IF} \sim 35$ (tetragonal HfO$_2$ phase), $k_{FE} \sim 30$).

Analyzing equation 6.6., it can be seen that the phase transformation in the interfacial paraelectric phase during cycling seen in the TEM study would lead to a decrease of the depolarization field. Figure 6.5c show the evolution of the ratio of depolarization and coercive field with cycling together with the trends of P_r and P_{r_rel}. An exponential-like drop of E_{dep}/E_c can be expected. As long as E_{dep} is in the same order as E_c, the difference of $P_{r,rel}$ and P_r is significant. Plotting the depolarization field-coercive field ratio evolution with cycling together with P_r and P_{r_rel} (see Figure 6.5c) indicate that as long as the ratio of E_{dep}/E_c is higher than one a decrease in relaxed P_r over time is visible. The decrease of E_{dep} is correlated within decrease of the ratio between the P_r and P_{r_rel}.

To complete the observations and study of the wake-up behavior in binary oxides, major findings of impedance spectroscopy study by Grimley et al.[167] were included. The authors reported that capacitance of the bulk region of the FeCap is constant until the onset of the degradation of the device. On the other side, the initial slight increase of the interface resistance during the wake-up is followed by the resistance drop of two orders of magnitude during the fatigue stage. Aforesaid performance hints at a reduction of the first slight decrease of number of defects that is followed with significant increase during the fatigue.

Before starting the modeling, some trends can be isolated from the electrical and structural characterization (Figure 6.6). Initially, i.e. during the wake-up stage, the polarization response increases, bias fields diminish whereas the leakage current stays nearly constant. From the structural perspective, the cycled FeCap device undergoes a phase transformation of M→O and T→O in the bulk and interfaces, respectively. Both the increase of the polarization response and the phase transition are accompanied by an increase of the total switching charge within the material system and continuous drop of the capacitance/k-value. In addition to

---

[15] Estimation of the depolarization field was performed based on the interface thickness and percentage of the M, T, O phase portions reported detailed TEM study performed on the same Gd:HfO$_2$ FeCaps [167].

6 FIELD CYCLING OF THE FERROELECTRIC HIGH-K MATERIALS

a continuous drop of the k-value, Figure 6.6a shows that the trend of amount of the charge in the system (ratio of the peak value-C_{max} and saturated capacitance-C_{min}) matches well with the evolution of remnant polarization. With further field cycling, the onset of fatigue (drop of the polarization) is visible. This is accompanied by an increase of the leakage current indicating a degradation of the stack, further phase transitions (even if not as prominent as in the beginning) and a continuing drop of the total capacitance and k-value as well [22].

Mechanisms

Based on the extensive characterization, a potential interaction among mechanisms affecting both

  a) dielectric (i.e defect – O vacancy and O ion – creation, diffusion, recombination) and
  b) ferroelectric (change of the remnant polarization by e.g. domain wall pinning, seed inhibition and other modifications of the switching process) properties

is derived as shown in the flow chart of Figure 6.6b.

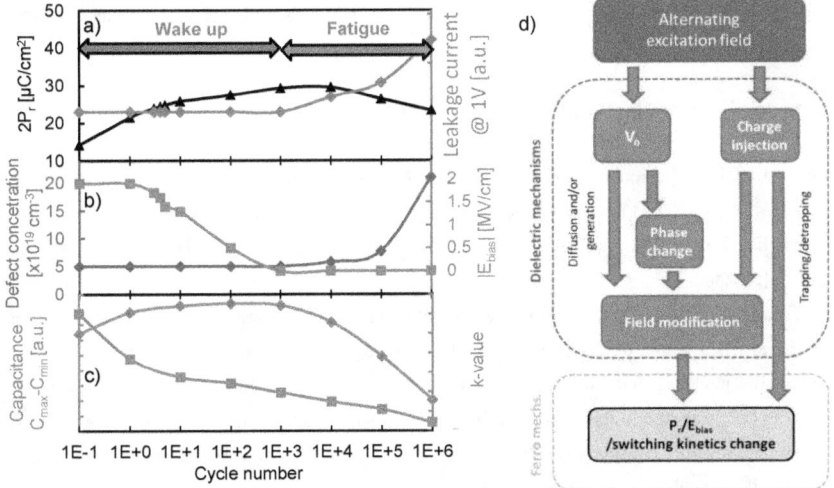

Figure 6.6. a) Evolution of the ferroelectric remnant polarization and the leakage current during bipolar cycling measured on Sr:HfO$_2$-based ferroelectric capacitors. b) Extracted defect concentration and bias field evolution as well as. c) Capacitance ratio and k-value cycling behavior. d) Flow chart of the device lifetime mechanisms. In the red box external excitation is given. The blue boxes depict mechanisms which result from this external stimulus and affect the dielectric properties. These mechanisms cause two separate processes which either promote or counteract the effective ferroelectric properties of the film stack (green box). Figure taken from Pešić et al.[22].

The mechanisms are connected to each other and take place at both interfacial and in bulk regions. Therefore, they cannot be perfectly separated in reality. For instance: the ferroelectric switching degradation might be connected to defect, i.e. O vacancy creation and subsequent charge trapping. The defect creation depends on the field, and increases with increasing field due to the switching and ion displacement of the ferroelectric material during cycling [21]. As it can be seen from Figures 6.6, the relation of all mechanisms in the stack is of rather complex nature necessitating an in-depth modeling study.

As mentioned before, like every high-k material [168], Sr:HfO$_2$ possesses a certain amount of oxygen vacancies within the material. The application of the alternating bipolar cycling voltage represents the main source of stress. The very field cycling causes the injection of charges into existing vacancies, which can change the polarization state of the ferroelectric material via local field modifications [154], as well as through the domain de-/pinning. This can result in either a rise or a deterioration of the P_r magnitude. In addition, the applied field can either generate new O/Hf vacancies or force the existing ones to diffuse through the material. As a consequence of the vacancy diffusion and generation, a fraction of the device volume can undergo a phase transformation. Here, transformation into switching or non-switching phase fractions together with the k-value change can either impede or facilitate the ferroelectric response by changing the local internal bias field. In the following section, the governing physical mechanisms will be assessed in detail.

6.2 Modeling of the Wake-up Behavior

In this section, investigation and in-depth modeling of the physical mechanisms responsible for the wake-up stage will be presented thorough a modeling approach. This approach relies on a combination of the dielectric-based simulation software MDLab [38,135] and the commercial TCAD Sentaurus device simulator [54] tool.

In order to reach better understanding of the correlation between ferroelectric and dielectric properties of the material listed within the last paragraphs of Section 6.1 and to unveil reasons behind the observed behavior, the modeling framework developed within Section 5.2. is further modified. The modeling framework comprises the commercially available MDLab software, used to investigate the leakage current and O vacancy defect creation and diffusion [158] and the TCAD Sentaurus device simulator, used to investigate the ferroelectric properties of the stack.

As discussed in the previous chapter, the wake-up behavior of the ferroelectric memory is attributed to the progressive decrease of the local internal bias fields. This decrease could be due to an O vacancy defect redistribution that can change their charge state, and/or induce partial phase transitions and k-value changes within the layer. Before simulating the wake-up and ferroelectric-dielectric interplay, the charge transport models developed in sections 5.2.3 and 5.2.4 are extended to account for the film properties described in following.

6 FIELD CYCLING OF THE FERROELECTRIC HIGH-K MATERIALS

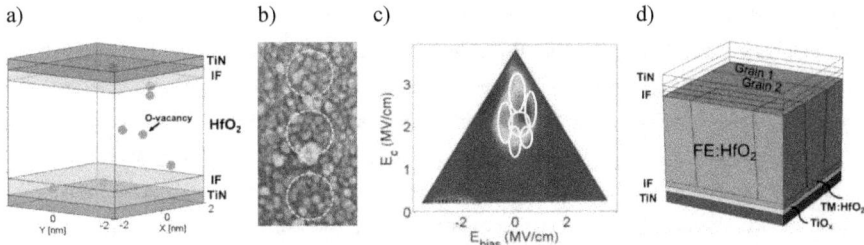

Figure 6.7. a) 3D FeCap device structure implemented within MDLab software. b) SEM picture of the polycrystalline hafnia: Granular structure of the capacitor device where each FeCap (circled area) has an arbitrary grain distribution. c) Corresponding domain separation depicted on a Preisach density plot. (d) 3D grain boundary TCAD model for accounting different grain and domain properties due to the polycrystalline nature of the film. Figure a) and d) taken from Pešić et al.[22].

First, to accurately account for the real characteristics of the FeCap device and phase transition behavior with field cycling, the impact of the two additional interfacial regions on the electrical behavior of the TiN/Sr:HfO$_2$/TiN capacitor has to be considered. Besides the mentioned TiO$_x$N$_y$ interfacial layer formation due oxygen scavenging from the HfO$_2$ and nitrogen diffusion into it, TEM study (see Figure 6.4) revealed parasitic tetragonal portions of the HfO$_2$ film towards the electrodes. Due to their tetragonal nature, these regions consist of non-switching transitional material (TM-HfO$_2$) (Figure 6.7) that is parasitically formed during the deposition of the HfO$_x$ and the subsequent annealing step [21,22]. Therefore, the complete device stack consists of TiO$_x$N$_y$/TM-HfO$_x$/FE-HfO$_2$/TM-HfO$_x$/TiO$_x$ sandwiched between two TiN electrodes, as depicted in Figures 6.7a and 6.7d. Due to the material intermixing that leads to formation of a thin ~1 nm thick sub-stoichiometric TiO$_x$ and TiO$_x$N$_y$ regions, which represent oxidized TiN and may affect the overall k-value of the stack. In addition to the lower k-value, these interfacial regions are characterized by a large defect density (mainly O vacancies). These, sub-stochiometric low-k TiO$_x$ regions are accompanied by a rough 1.5±1 nm thick HfO$_2$ interface predominantly with a higher k-value due to the presence of a tetragonal phase as seen in TEM study (Figure 6.4b). These two factors significantly impact the voltage distribution inside the dielectric layer: Hence, a superposition of the lower dielectric constant and the electron trapping at interfacial defects can results in either a local increase or decrease of the field, which affects the resulting polarization of the ferroelectric film.

Further, the polycrystallinity of the HfO$_2$ layer has to be explicitly taken into account. As mentioned in Chapter 3.1, Sr doped HfO$_2$ was crystallized using an 800 °C for 20 seconds anneal in nitrogen atmosphere causing the granular morphology of the film. Scanning electron microscopy of the bare HfO$_2$ oxide film showed that certain device comprises random grain sizes (Figure 6.7b). As reported by Mulaosmanovic et al. [169], these grains can be correlated with discrete ferroelectric domains and switching events within highly scaled FeFETs. Therefore, it can be expected that the total switching response e.g. a Preisach switching density (see Figure 6.7c) of the FeCap consists of a group of grains/domains with slightly varying properties. To address the polycrystalline nature and variability of the film properties, the modelled hafnia ferroelectric layer is divided into randomly distributed grains and two interface layers next to the electrodes. The granular morphology and grain boundaries play very important role since they tend to act as the preferable locations for the accumulation of the oxygen vacancies. As discussed in Chapters 2 and 5, oxygen vacancies actively participate in the trap-assisted current and are responsible for the local field modifications due to charge trapping. The total area of the measured capacitor (33000 µm²) consists of thousands of grains which all might have a slightly different domain orientation, coercive field or remnant polarization. In order to model the variability in the resulting switching characteristics, each of the five modelled grains represents an averaged ensemble of domains with similar properties. Each ensemble is defined by a distinct coercive field and remnant polarization value. The coercive fields used for the ensembles were chosen in the range of 1.6 MV/cm to 2.9 MV/cm to be consistent with the FORC results in Figure 6.7c, whereas based on the P-V characterization the remnant polarization of the domains was distributed between 15 µC/cm² and 24 µC/cm².

To address the phase transition with field cycling, different phases of the doped hafnia, which can coexist inside the same grain: a) the orthorhombic phase, responsible for the ferroelectric switching; b) monoclinic and c) tetragonal phases, which are not active from the ferroelectric point of view were taken into account. Further, charge transport within the model was calibrated in Chapter 5. The extracted defect distributions were fed into the 3D grain boundary model of the FeCap developed within the Sentaurus device simulator to simulate the ferroelectric response.

In order to access the possible material diffusion due to the application of high electric fields, the diffusion and recombination of oxygen ions and vacancies were calculated through a kinetic Monte Carlo (kMC) model implemented in the MDLab package [135]. Diffusion rate R_D and recombination rate R_R are calculated using the equation 6.7a and equation 6.7b, respectively:

$$R_D = v \cdot exp\left(-\frac{E_{A,D} - Q\frac{\lambda}{2}E_{eff}}{k_B T}\right); \quad (6.7a) \qquad R_R = v \cdot exp\left(-\frac{E_{A,R}}{k_B T}\right) \quad (6.7b)$$

, where Q is the charge of the diffusing space, v Debye vibration frequency, λ is the jump distance which is typically 3Å, and $E_{A,D}$ is the activation energy for ion/vacancy diffusion, E_{eff} the effective field along the jump direction, $E_{A,D}$ the activation energy (energy barrier) given by $E_{A,D}$=0.7 eV and $E_{A,D}$=1.5 eV for O_2^- ions and positive O vacancies ([38,158]), respectively. $E_{A,R}$ represents the activation energy for recombination between complementary species.

Recombination and diffusion processes were coupled to the charge transport model developed before. From the equation 6.7a it can be seen that beside the activation energy $E_{A,D}$, the calculated diffusion rate strongly

6 FIELD CYCLING OF THE FERROELECTRIC HIGH-K MATERIALS

depend on the applied field E_{eff}, temperature and film stoichiometry. The whole diffusion process is illustrated in Figure 6.8.

Finally, to emulate the stress-induced defect generation, the thermochemical bond breakage model was implemented in the MDLab software [136, 172, 176]. Hafnium oxygen (Hf-O) bond breakage generates an oxygen vacancy, which is an electrically active defect contributing to the TAT transport, and an oxygen ion. The vacancy generation rate G is given by equation 6.8.

$$G = v \cdot exp\left(-\frac{E_{A,G}-bE}{k_BT}\right) \qquad (6.8)$$

, where v is the bond vibration frequency, b the bond polarization factor, E the electric field, $E_{A,G}$ the activation energy for generation of ion-vacancy pair, k_B the Boltzmann' constant and T temperature.

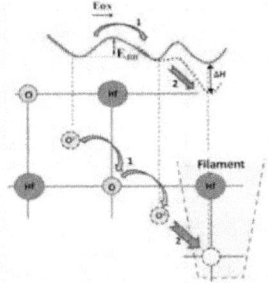

Figure 6.8. Sketch of the energy profile along the lowest-energy pathway for long-range diffusion of doubly charged oxygen vacancy in monoclinic HfO$_2$. Figure 6.8a reproduced according to [158];

The same model was also used for the simulation of the leakage current evolution with cycling (see Figure 6.1c), which allows extracting the increase in the density of O vacancy defects within the stack. The increase of the vacancy density seems correlated with the P_r evolution observed with cycling, (Figure 6.6a). However, it should be noted that the density of defects evaluated from the leakage current corresponds mainly to those located at the grain boundaries, where they account for the main contribution to leakage current [136]. More precisely, the leakage current thought the grain boundaries is more than tenfold higher than through the grain itself. Thus, the diffusion of the vacancies thought the grain would keep the leakage current constant during the wake-up stage. The role of O vacancy/ion diffusion is also consistent with the fact that HfO$_2$ based devices are known for their high oxygen mobility [155]. This claim is strengthened by the fact that the electric field applied on the HfO$_2$-based ferroelectrics is at least one order of magnitude higher than the fields typically applied to PZT [155]. Furthermore, a recent transmission electron microscopy study confirmed the generation and movement of oxygen ions and vacancies within 10 nm thick HfO$_2$ film under similar operating conditions [170]. Starschich *et al.* just recently proved resistive and ferroelectric switching operation in the same Y:HfO$_2$-based device[171]. It is widely accepted that oxygen vacancy motion is required for resistive switching in HfO$_2$[172].

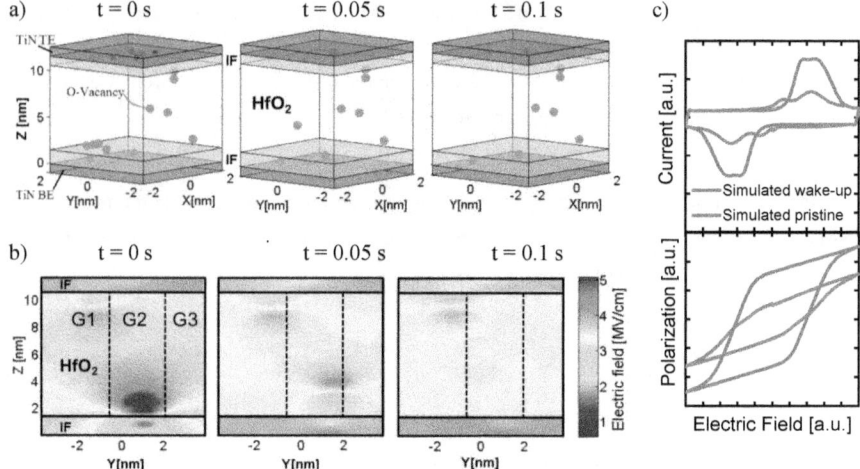

Figure 6.9 Simulated wake-up of the TiN/IF/Sr:HfO$_2$/IF/TiN device: a) Vacancy (red dots) evolution with the field cycling of the MIM FeCaP in 3 different stages (in the beginning, t=0s, mid of wake-up, t=0.05s and in woken-up stage, t=0.1s)[16]. b) Corresponding electric field evolution within the device with the field cycling of the FeCap in three different points in time at 4 MV/cm external applied field. IF represents a low-k interfacial layer whereas G1-G3 represent three different grains within the stack. c) Resulting I-V and P-V characteristics obtained by removing the charges from the interface and changing the k-value of the grains undergoing a phase transformation. Figure was taken from study by Pešić et al.[22].

In order to examine the cycling influence and dynamics within the stack as well as its influence on the field distribution (predominantly on internal bias fields), the diffusion and recombination of oxygen ions and vacancies through a kinetic Monte Carlo model implemented in the MDLab package [135] was simulated. Diffusion rates (equation 6.7a) are calculated by considering activation energy for ion and vacancy diffusion of 0.7 and 1.1 eV respectively [170, 173]. Based on the previous discussion the pristine state constellation within the model was set so that the transitional regions at the dielectric interface have much higher defect density. The so-called local internal bias field was created by highest density of the vacancies in the interfacial regions (Figure 6.9a). This scenario was used to simulate the preferable movement of the O vacancies driven by the applied electric field. Simulation of bipolar stress cycling revealed that O vacancies redistribute uniformly within the grain (Figure 6.9a). In addition to this drift/diffusion process,

---

[16] For the reasons of clarity and simplicity, the depicted model shows just a few vacancies within the device. To better illustrate the mobility of vacancies (red dots) due to the applied field a right-hand side of the device was "locked" setting the infinite numbers for activation energy needed for vacancy diffusion.

recombination can occur. Due to the recombination of vacancies and interstitial ions the internal bias field decreases. A more uniform field distribution (Figure 6.9b) is created within the stack, resulting in homogeneous switching of all domains within the device. In order to confirm the influence of the temperature observed in the temperature dependent wake-up characterization, the simulation was repeated for 150 °C. With increasing temperature, the vacancy diffusion is accelerated resulting in more mobile vacancies and faster wake-up, which is in agreement with results presented in Figure 6.3a-b. The homogeneous field distribution was achieved two orders of magnitude faster with respect to room temperature, which agrees very well with the experimental results shown in Figure 6.3a-c where 10^3 cycles at room temperature and two cycles at 150 °C were needed to reach the identical, completely woken-up state.

Besides the more homogenous field distribution due to vacancy diffusion, the movement of ions can cause possibly the phase transformation. *Ab-initio* calculations [174] showed that incorporation of the oxygen vacancies into the lattice can strongly influence the phase stability of the HfO_2. They showed that dopants and also vacancies favor ferroelectric and tetragonal phase, i.e. decrease their formation energy compared to the monoclinic reference phase (Figure 6.10). This is confirmed also by TEM results that show the transition of the initially tetragonal phase of non-switching $TM-HfO_2$ interface to orthorhombic switching) phase for the woken-up device. A thickness reduction of this dead layer [153] at the interface increases the ferroelectric switching response of the device. Thus, the remnant polarization increases during the wake-up.

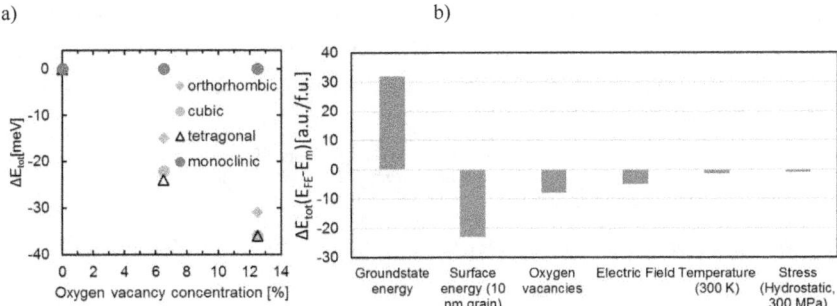

Figure 6.10. a) *Ab-initio* calculated total energy change of the monoclinic, orthorhombic, tetragonal and cubic phases of HfO_2 as function of the oxygen vacancy concentration. b) Influence of different parameters on the energetic difference between ferroelectric and monoclinic reference phase. Figure taken and reproduced from [174] and [69].

It should be noted that even though the vacancies move towards the middle of the device, the leakage current stays constant because it is dominated by the leakage through the grain boundaries and not through

6 FIELD CYCLING OF THE FERROELECTRIC HIGH-K MATERIALS

the grain volume. Therefore, vacancies that move during the wake-up phase indeed increase the leakage current component through the grains, however this is insignificant compared to the total current.

The trap concentrations and distributions obtained from compact modeling of the diffusion processes were fed into a 3D ferroelectric capacitor model developed in TCAD (see Chapter 5)[17]. In order to assess the ferroelectric switching, modeling is indispensable to obtain a realistic influence of charge and field on the switching of the device. To model the behavior of a ferroelectric, the model has to account for the history dependent charge-voltage relationship of the ferroelectric as discussed in Chapter 2. Based on the FORC results and ferroelectric characterization, the properties of the modeled ferroelectric material are tuned in order to obtain the desired characteristics.

The initial stack is also affected by the presence of defects, whose charge can unpredictably alter the local field, affecting the ferroelectric switching performance. To account for this effect, it was considered that oxygen vacancies are preferentially distributed at the interfacial regions. These regions are mainly made of tetragonal phase HfO_x, with higher permittivity, which additionally affects the local field.

Further, these regions are considered non-switching in the pristine state, i.e. they act as a passive (dead) layer [153]. Furthermore, monoclinic (non-switching phase) was modeled setting the grains to a low-k non-switching state. Using the Preisach model described in section 2.3, a double peak I-V characteristic as well as a pinched hysteresis loop was successfully simulated (see Figure 6.9c). Domain de-pinning was represented by removing the charges ($2 \cdot 10^{19}$ cm^{-3}) within the interface, whereas the phase transformation was included by changing of the respective k-values[18] and setting the portions of interface as well as the previously non-switching grains to switching state. These modifications resulted in merging of the peaks and opening of the hysteresis curve (see Figure 6.9c), which completely emulates the measured device behavior. The parameters and the influence of the simulated wake-up are summarized in Table 6.2 and compared with the measured values.

Table 6.2 Comparison of the simulated and measured k-value change and its influence on the wake-up.

	k-value change [%]	P_r change[19] [%]
Simulated	3*	44
Measured	4	51

*Calculated/modeled for the values given in Table 6.1.

---

[17] In contrast to the model implemented in a MDLab software, TCAD simulator does not allow random distribution of defects. Therefore, this randomization was mimicked via the definition of the regions within the TCAD model (i.e. top left corner, central region of hafnia, etc.) and its equivalent in the MDLab software. Further, the extracted defect concentration (from the MDLab) and its spatial location was transferred to TCAD by: 1) identifying equivalent regions of the MDLab and TCAD model and 2) transferring the normalized concentrations of stage before and after the cycling from MDLab software into the designated portion of TCAD device.

[18] To address the phase transition k-values for T, O and M-phase given in Table 6.1 were used.

[19] It should be noted that P_r change is not only due to the k-value change but also to the fact that these previously non-switching regions start switching and contribute to the resulting switching current.

6.3 Polarization Fatigue and Dielectric Degradation

After an in-depth study of the effects dominating the so-called wake-up, responsible for enhancement of the ferroelectric properties during the initial cycles, the subsequent gradual degradation of these properties is studied. Similar to the wake-up stage leakage current defect spectroscopy [96,175,21] was used to examine the quality of dielectric in terms of defects generated/activated within the dielectric layer during the cycling stress. In parallel to the endurance behavior, the evolution of the static DC leakage current with the bipolar cycling was monitored.

To investigate the influence of the field cycling on the device properties in detail, pristine capacitors where preconditioned with 10^3 bipolar cycles to establish a fully woken-up state. The endurance of these devices stressed with unipolar and bipolar pulses is given in Figure 6.11a. It can be seen that only the alternating switching, i.e. a continuous change of the polarization state results in a degradation and consequent reduction of the MW. On the other side, the unipolar stress does not influence the MW significantly. Leakage current defect spectroscopy shows that independent of the polarity of the unipolar stress pulses, both leakage current and memory window stayed constant. Hence, it is reasonable to assume that alternating polarization switching itself (continuous ion displacement within the ferroelectric crystal lattice) induces the endurance degradation and fatigue in contrast to the unipolar, non-switching stress. In Figures 6.6a and 6.11c, a direct correlation between the increase of the defect concentration and the degradation of the MW during the fatigue stage can be seen.

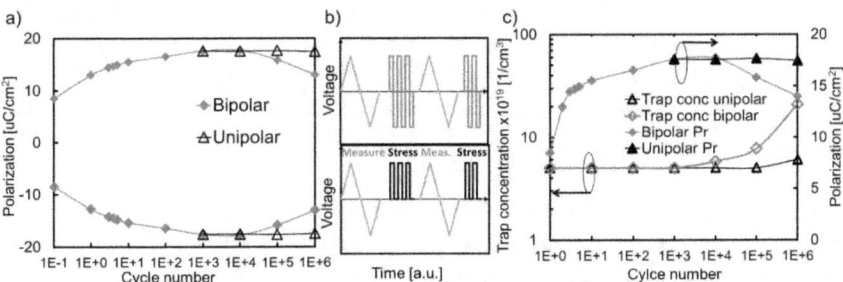

Figure 6.11 a) Fatigue characteristics of a Sr:HfO$_2$ based capacitor measured after the full wake-up was achieved, the device was stressed with unipolar (black) and bipolar (red) stress at 10 kHz stress frequency. b) Pulse sequence obtained for the recording of the endurance characteristics. For the reasons of clarity and brevity negative unipolar pulses were omitted from the figure. c) Defect concentration evolution (extracted at 1 V) with unipolar (black) and bipolar (red) cycling. Figure was taken from study by Pešić et al. [21].

To address the observed behavior and investigate the mechanism behind it, the thermochemical bond breakage model was implemented in the MDLab software [136,176] as discussed above (see equation 6.8.).

6 FIELD CYCLING OF THE FERROELECTRIC HIGH-K MATERIALS

Equation 6.8 indicates that the bond breakage heavily depends on the applied field and the activation energy, which is usually given by the stoichiometry of the film. In refs. [21] and [22] it was reported that faster O vacancy defect generation occurs within the TiO_x interface next to the electrodes. These sub-stoichiometric, non-switching regions are characterized by a high number of defects and a lower permittivity (higher field drop), resulting in an increased factor of degradation and causing faster bond breakage. The defect generation and the subsequent charge trapping significantly influences the local field distribution over the device stack. The field over the interface increases, while it reduces inside the ferroelectric bulk [21]. Hence the ferroelectric layer (switching/active part of the device) experiences a lower field, which leads to a reduced number of switching domains if the coercive field of some domains is not reached anymore. All these effects cause a decrease of the memory window. In contrast to the wake-up stage, the continuous cycling is not only characterized by vacancy diffusion, but also by a vacancy generation within the grains. With increasing stress and corresponding degradation, these defect locations start representing a significant leakage current path besides the previously dominating leakage through the grain boundaries.

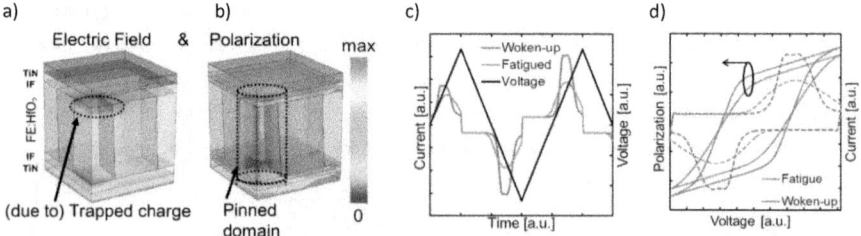

Figure 6.12 Charge trapping influence on the electric field: Polarization response and corresponding domain pinning due to the charge trapping of a) woken-up and b) fatigued stack. For the reasons of clarity single domain pinning is illustrated. Interface, IF comprises parasitically grown TiO_x as well as non-switching portion of HfO_2. Comparison of the woken up and degraded state obtained by simulation: c) Current and voltage transients. (d) Polarization and current voltage response. A broadening of the switching peak and consequent decrease of the remnant polarization is visible. Figure c) and d) are taken from study by Pešić et al. [21].

In order to solidify this assumption, the modeling of the degradation and its influence on the ferroelectric switching was performed. Extracted defect concentration and obtained defect distribution were fed into the 3D grain boundary model of the MFM capacitor that was developed within Sections 5.2.2. and 6.2. The obtained result (Figure 6.12) verified the degrading influence on the field distribution across the stack and the resulting current voltage and polarization voltage characteristic (Figure 6.12c-d). Besides the pure electrostatic influence of shielding, the charge trapping creates dipoles impeding the switching of the domains, which results in a partial or even complete pinning of the domains [21]. For sake of simplicity, the simulated example given with Figure 6.12 a-b shows charge trapping within one corner of the top interface region, which alters the local field distribution and pins the domain in the encircled grain. As in

6 FIELD CYCLING OF THE FERROELECTRIC HIGH-K MATERIALS

the measurement, also the simulated current-voltage traces resulted in broadened peaks with decreasing magnitude. Consequently, the simulated polarization hysteresis was characterized by smoother transitions and lower opening.

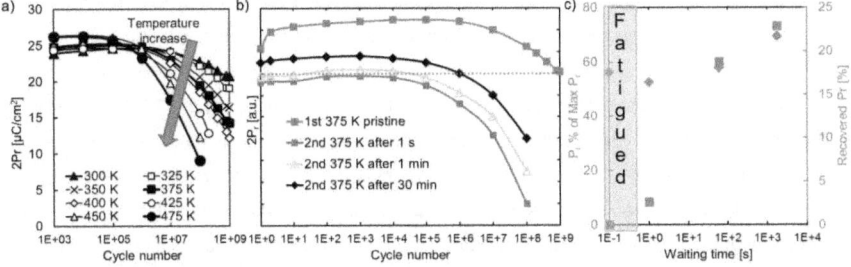

Figure 6.13 a) Memory window evolution as function of device cycling and temperature. Increased temperature results in an earlier fatigue onset. Higher P_r values were extracted with increasing temperatures due to the parasitic leakage current increase which affects the $2P_r$ extraction procedure. b) Polarization recovery endurance characteristics obtained (after the initially recorded endurance sequence of 10^9) for three different waiting times indicating the influence of the de-trapping and relaxation processes. c) Polarization recovery with respect to the fatigued state (marked with green). Figures a) - c) are taken from study by Pešić et al. [21][22].

To further substantiate the O vacancy defect generation and consequent charge trapping as the root cause for the memory window degradation, temperature dependent experiments were performed. Firstly, temperature-dependent endurance was recorded (Figure 6.13a). Temperature dependent endurances revealed an earlier onset of fatigue as well as enhanced fatigue with increasing temperature (Figure 6.13a). If the trapping represents a component which strongly influences a field distribution, switching properties and resulting memory window, it is reasonable to expect that de-trapping would lead to restoration of the memory window. Hence, recovery experiment was performed in order to fully assess the influence of the trapping and stack degradation (see Figure 6.13b-c) [21]. The experiment was performed as follows: After the completion of the endurance test, a second endurance experiment was performed. To investigate the time scale of the de-trapping process, waiting time between the two endurance measurements was varied. First, several devices were stressed with 10^9 bipolar cycles. Subsequently, each of the devices was stressed again with the same amount of cycles with progressively increasing waiting time between the runs. The first experiment was carried out in such a manner that the second endurance was recorded immediately after the first run. To provide the same initial condition after the reference cycling, the next already preconditioned device was stressed after a sequentially increased waiting time. As indicated by both modeling and experiments performed before, trapping and degradation had a strong influence on the ferroelectric switching and memory characteristic of the ferroelectric device. With increasing waiting time, a stronger recovery of the remnant polarization was observed. This behavior clearly corresponds to accelerated de-trapping of electrons from occupied defects. According to this explanation, the longer the waiting time, the more defects were de-trapped, resulting in domain de-pinning, recovery of the polarization state and re-opening of the memory window. Besides the opening of the memory window, recovery experiments showed an earlier onset of the fatigue (with respect to the first endurance run) for all devices independently of the waiting time. This behavior confirms the previously discussed degradation of the layer. To complete the observations and study of the degradation and fatigue behavior in ferroelectric, a discussion of the results on the hot atom degradation as reported by Masuduzzaman et al. [177,178] is given. The authors reported a hot atom overshoot being responsible for the degradation and bond breakage due to the alternating voltage stress. Even though the authors proposed a very straightforward and reasonable explanation, the concept should not be called hot atom but hot ion degradation due to the fact that a displacement of the ion stresses the lattice (Figure 6.14a). Regardless of the name, an overshoot of the ion can be caused if pulses with ultra-steep flanks are applied. When changing the polarization state, the ion has a kinetic energy high enough, not just to overcome the barrier but also to climb to the opposite wall of the energy potential eventually reaching the bond breakage energy. This bond breakage results in the generation of the defects, i.e. O vacancies and interstitial O ions in this case, which finally lead to dielectric breakdown. To examine the applicability of the concept, endurance experiments were performed. The plateau width of the pulses was kept constant (1 μs) whereas the flanks of the PRG/ERS stress pulses were varied from 100 ns to 1 ms (Figure 6.14b, inset). Indeed, the increase of the rise/fall time of the stress pulses resulted in better endurances. Even though the experiment was successfully repeated, a detailed statistical approach has to be performed in order to draw a final conclusion on/about the applicability of the method.

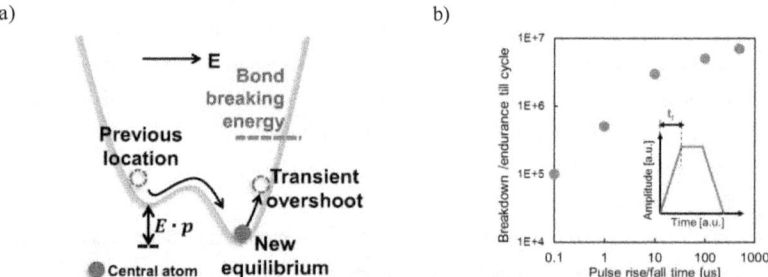

Figure 6.14 Hot ion degradation concept. a) The double-well energy potential for each switching event, b) Breakdown-endurance dependence on the rise/fall time of the stress pulse. (inset) Pulse with variable rising flank. Figure a) reproduced from [177][178].

Finally, based on the combination of the results of electrical characterization and modeling a physical model of the mechanisms governing the wake-up and fatigue effect within the FeCap can be depicted (See Figure 6.15). Pešić et al. [22] reported that the mechanism responsible for the wake-up behavior are

6 FIELD CYCLING OF THE FERROELECTRIC HIGH-K MATERIALS

material and charge drift /diffusion due to the alternating field cycling as well as structural changes of the material (T→O and M→O phase transition). In parallel, also the mechanism behind the polarization fatigue and degradation of the dielectric take place, but it starts to dominate just at higher numbers of switching cycles. The generation of new defects (oxygen vacancies) and injection of charges modify the switching properties of the device giving rise to the pinning of domains and a consequent reduction of P_r.

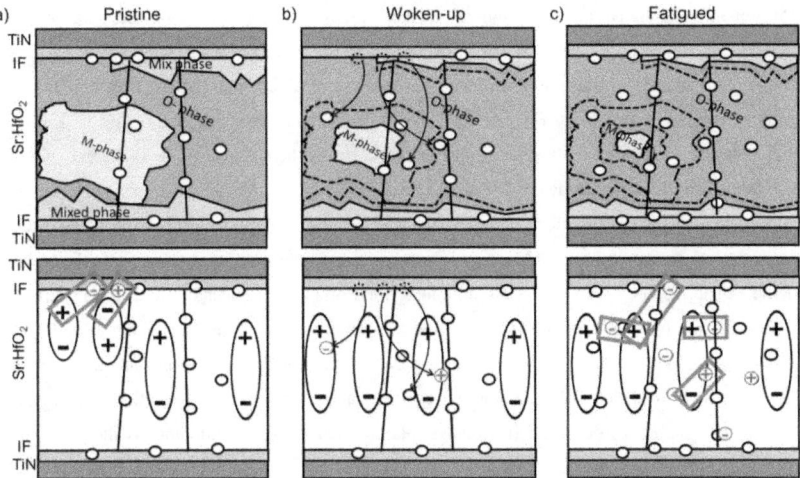

Figure 6.15. The schematic illustration of the structural changes and corresponding physical effects ongoing during the three different stages of the FeCap lifetime a) pristine, b) wake-up and c) fatigued. Pinned domains are represented with the red rectangles; circles (solid line) denote the O vacancies, circles (dashed lines) a previous position of the O vacancies and circles with + an – sign denote positively and negatively charged vacancies, respectively. Figure reproduced based on study by Pešić et al. [22].

6.4 Road Towards 1T Ferroelectric Memory-FeFET

After the comprehensive study of field cycling behavior of the FeCap that resulted in the development of a model being able to emulate the physical mechanism within the device and its lifetime, in the following this model will be extended to FeMIS and FeFET structures. It is important to mention that the integrated test structures that were used for characterization screening were not just simple MIS-capacitors but were transistor structures with source- and drain-regions connected to a metal layer. As discussed at the introduction, independent of the memory cell architecture (1T-1C or 1T), hafnia-based ferroelectric memories suffer from limited endurance. Both FeMIS and FeFET models will be developed. However, the main focus of this chapter will be on the comparison of the integrated FeMIS capacitor multi-structures with the previously analyzed lab-scale FeCap devices.

6.4.1 Electric Field Cycling Behavior of FeFET

The properties of the FeMIS and FeFET structures were investigated similar to the FeCap characterization. This characterization was carried out on integrated multi-structure of DUTs (2000 devices in parallel, sharing the gate and bulk terminal). The multi-structure DUTs were chosen to achieve a sufficient device area for obtaining a signal large enough to measure area-scaled properties of the gate stack (e.g. polarization, capacitance). Besides the sufficient area needed for distinct parameter study, a multi-structure consisting of a large number of small devices enables studying small channel effects (even though they are averaged) in contrast to one large FET device. Due to the polycrystalline doped-hafnia utilized as the ferroelectric within the gate stack, it is reasonable to expect similar behavior seen like in the FeCap. Indeed, the characteristics of the multi-structure FeFET resulted in a double current peak I-V that ended up merged into a single-peak characteristic after a certain initial cycling (Figure 6.17a). This peak merging caused an opening of the polarization hysteresis loop as explained in the chapters before. In contrast to the endurance plot of the FeCap, Figure 6.17b shows a continuous trend of P_r vs. number of switching cycles for the MFIS. Even woken-up (10^3 cycles) a leakage charge continuously increases, especially after 10^3 cycles when it starts to represents a significant influence on the total polarization (integral of) charge (see 6.17b inset). In order to observe the pure polarization characteristics a PUND measurement (see Chapter 4.1) method can be applied. However, the non-compensated endurance is shown in order to depict the influence of the intrinsic operation of the device.

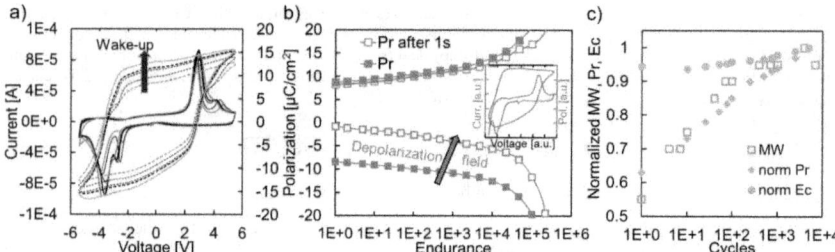

Figure 6.17 Field cycling behavior of the integrated FeFET device. a) I-V and P-V characteristics of the FeFET structure measured with triangular waveform with amplitude of 5.5 V and pulse edges of 1 µs. b) Field cycling evolution of the remnant polarization P_r (closed symbols) and relaxed remnant polarization measured after 1 s (open symbols). The label "depolarization field" indicates the cause for this relaxation. The inset shows a leaky characteristic of the degraded FeFET device after 10^3 cycles. c) Field cycling dependence of the normalized MW and corresponding P_r and E_c evolution. Figure (c) reported in [179].

In addition to the P_r measured directly, the so-called relaxed polarization [142], i.e. the measured P_r after one second delay was recorded. In Figure 6.17b, the influence of the depolarization and the de-trapping can

6 FIELD CYCLING OF THE FERROELECTRIC HIGH-K MATERIALS

be seen. A strong relaxation of the negative remnant polarization occurs. The depolarization field E_{dep} represents one of the biggest issues of the FeFET concept [166] and, as can be seen, directly influences the stability of the polarization state. This perturbing effect can be calculated as:

$$E_{dep} = -P\left(\varepsilon_0 k_{FE}\left(\frac{C_{IF\_sem}}{C_{FE}}\right)\right)^{-1} \quad (6.9)$$

, where P denotes the polarization, ε_0 the vacuum permittivity, k_{FE} permittivity of the ferroelectric film, C_{IF_sem} capacitance of the interface buffer layer and semiconductor combined and C_{FE} the capacitance of the ferroelectric.

In order to correlate the intrinsic polarization of the device with the evolution of MW and E_c with field cycling, PUND measurements were performed. In Figure 6.17c, a direct correlation between the remnant polarization and the increase of the memory window can be seen. However, the coercive field remains nearly constant with just a slight increase with field cycling. Returning to the first order approximation for MW calculation given by $2E_cd$ (defining MW as the product of two times the coercive field and thickness) can be seen that MW evolution with field cycling cannot be explained. Even though, this relation was reported by many studies, experimental results (see Figure 6.17c) revealed that beside the E_c and thickness, d, additional parameters have a much stronger influence on the FeFET's memory window and its evolution with field cycling. Moreover, the memory window showed direct dependence on the P_r value whereas the coercive field remained almost constant during the cycling.

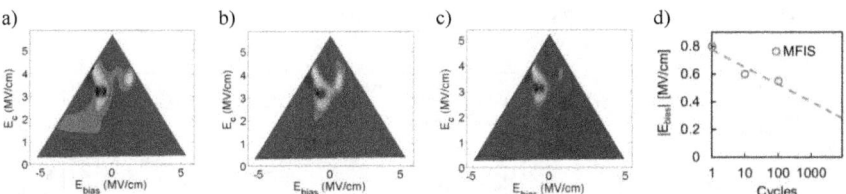

Figure 6.18 Evolution of the Preisach/switching density ρ during wake-up phase of MFIS device cycling after: a) 1; b) 10^2 and c) 10^3 cycles. d) Trend of the internal bias field obtained with reduction of the cycling amplitude.

In order to verify the observed trends and extract the internal bias fields, a Preisach/switching density evolution during MFIS device cycling was recorded (Figure 6.18). From the FORC measurements can be seen that like in the measurements of MFM, internal bias diminished, whereas the E_c remains constant with cycling. In the MFM, the local internal bias field completely disappears with cycling, whereas here (MFIS case) this is not the case. Due to the asymmetry of the stack, the internal bias remains even after the merging of the transient current peaks, which usually is defined as the end of wake-up. The devices could not be cycled to the fatigue regime, because the FORC measurement creates a high stress that caused an earlier hard dielectric breakdown. Therefore, both stress and monitoring amplitudes were reduced to address the internal bias field evolution in all device lifetime stages. However, since the screening voltage was reduced, only single peak was visible in the Preisach distribution (Figure 6.18c). Hence, only some trends instead of a complete picture can be derived (Figure 6,18d). From Figure 6.17a, a turning point in internal bias field evolution with cycling can be seen. After the initial decrease of the internal bias field, the merged switching current peaks shifts towards more negative fields. This points to the possibility that defects are generated and the resulting enhancement of charge trapping shifts the whole characteristics. An imprint occurs. As can be seen from equation 4.2, leakage current has a negligible impact on the FORC plots if it does not strongly change between the measured current branches of subsequent reversal fields.

As discussed before, a combination of the asymmetric stack, or electrodes with different workfunctions causes the shift of the complete *I-V* and corresponding *P-V* characteristics. Here, Si and TiN cause additional asymmetry due to their different nature (semiconducting vs. metallic). This asymmetry necessitates the use of high fields and causes a charging of the stack. Consequently, a complete closure of the MW can be seen before breakdown, whereas the transfer characteristics of the device remain measurable. To study the imprint of the MFIS stack, reduced operating voltages were chosen where device was driven within sub-cycles which reduces the total stress and prevents a breakdown of the device stack. In addition, Lab-scale MFIS (33000µm^2) devices with higher endurance strength with respect to the integrated hardware were used (see section 3.1 for details). Typical characteristics of the MFIS devices are given in Figure 6.19a and were recorded at -6/4V ERS/PRG pulses at 10 kHz.

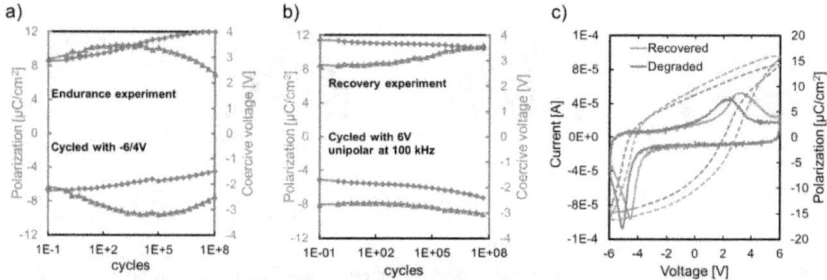

Figure 6.19. Endurances and imprint behavior due to the charge trapping with the MFIS structure. (a) Endurance characteristics of the Si:HfO$_2$ based lab-scale MFIS capacitor, tracing change of the P_r and E_c of the stack. b) recovery experiment performed by unipolar cycling inducing the de-trapping and an increase of the MW. c) Degraded and recovered *I-V* (solid line) and *P-V* (dashed line) characteristics.

Similar, to the FeCap device, an initial increase of the MW during the wake-up stage is followed with the decrease of the MW. In addition to the extracted remnant polarization, the E_c evolution with cycling is plotted. Interestingly, the coercive field is being shifted toward more positive values due to the charge

6 FIELD CYCLING OF THE FERROELECTRIC HIGH-K MATERIALS

trapping already at about 10^2 cycles. With cycling, this shift becomes so strong that most of the domains formerly contributing to the switching current cannot be accessed by the applied field anymore. The problem is similar to what was explained by Schenk et al. [108] for the pristine case in FeCaps, but here it is due to bias fields introduced by charging during cycling. Hence, the closure of the MW is observed as a pure consequence of the imprint. Next, the imprint behavior due to the charging and the recovery steps was studied. Same, already pre-cycled (imprinted) MFIS devices were used for the recovery experiments. Further, *I-V* and corresponding *P-V* characteristics of the imprinted MFIS device under test is given with Figures 6.19c.

To verify the hypothesis of trap charge being responsible for the imprint and closure of the MW within the MFIS stack recovery experiments were performed. By applying unipolar pulses (-6V) on the gate a back-shift of the coercive field and a recovery of remnant polarization and MW was seen. Comparison of the imprinted and recovered *I-V* and *P-V* characteristics is shown in Figures 6.19c. Here, a clear trend of the current peak shifting toward more negative values can be seen. Consequently, more domains can be accessed by the field amplitude and can start contributing to the switching current again, hence resulting in opening of the hysteresis loop (Figure 6.19c).

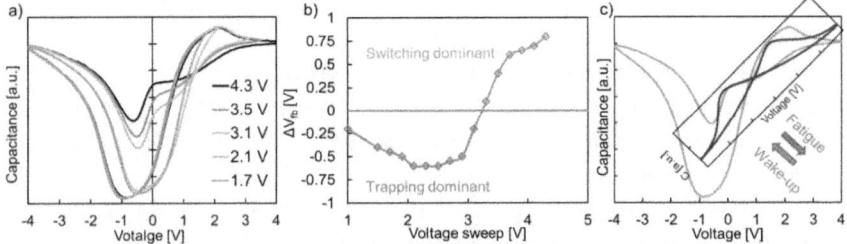

Figure 6.20. *C-V* characteristics of the MFIS capacitor structures. Amplitude dependence of the a) *C-V* characteristics and (b) extracted flatband voltage measured on MFIS integrated multi-structure. c) Correlation of the *C-V* characteristics of MFM and MFIS structures. The arrows denote increase/decrease of the peaks with wake-up and fatigue respectively.

As a part of the standard as well as the trapping characterization, *C-V* characteristics were recorded. The measurements were performed on Si:HfO$_2$ based integrated multi-structures with the 3.5 cat% Si dopant content. Keithley 4200 SCS Parameter Analyzer equipped with a Multi Frequency Capacitance Measurement Unit was utilized for the *C-V* characterization. *C-V* measurements were recorded using an AC frequency of 100 kHz, a small signal modulation with amplitude of 50 mV and a hysteretic DC voltage sweep starting from -4 V up to 4 V in steps of 0.1 V and back. In contrast to the standard MIS characteristics comprising accumulation, depletion and inversion region a *C-V* characteristic possess additional features which are the consequence of the interference of the charge trapping and ferroelectric switching (see Figure 6.20a). This interference can be explored by a sequential increase of the positive voltage sweep. In Figures 6.20a-b this dependence can be seen. A voltage amplitude of around 3.2 V is needed to backswitch the device into the low-V_{th} state. All other voltage amplitudes (lower than 3.2 V) resulted only in a pure trapping response, causing a shift of the high-V_{th} state to even more positive fields. In addition to the shift of the flatband voltage to the more negative fields, with backswitching a peak of the C-V characteristics around 2 V arises. This peak represents additional charge generated by ferroelectric switching with respect to the saturated value of the C-V characteristics. Moreover, together with the peak arising at around 0 V it forms a kinked butterfly C-V characteristic well known for the MFM devices. This is verified by the fact that analog to the MFM, C-V maxima of the C-V characteristics rise and diminish with wake-up and fatigue respectively.

6.4.2 Modeling of the FeMIS and FeFET Stack

In the following, a transition from MFM via metal-ferroelectric-semiconductor (MFS) to MFIS FeFET structures will be discussed and simulated. Furthermore, design rules for the MFIS FeFET structure are derived. Finally, a correlation of all determining parameters will be presented. All modeling and simulation approaches performed within this study have been carried out using a commercially available Synopsys TCAD framework.

Based on the detailed characterization of FeCap structures and analogous behavior of both the FeMIS and the FeFET stacks to these FeCap structures, device models for FeMIS and FeFET were developed. Stack of the previously derived FeCap model was placed on the interfacial buffer layer (SiO$_2$) and a Si p-type substrate instead of a TiN bottom electrode. Similar to the FeCap, the history dependent behavior of the ferroelectric switching was accounted utilizing the Preisach-based hysteresis model available within Sentaurus Device simulator as described in detail in Section 2.3.

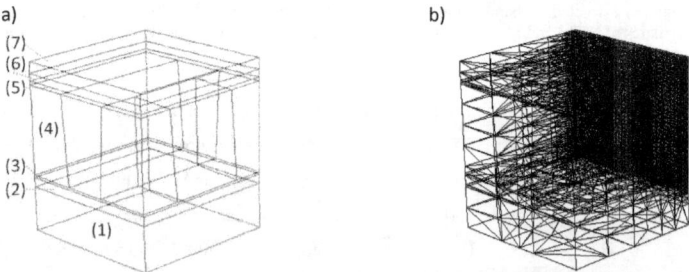

Figure 6.21 Geometry of the developed 3D grain boundary model of the FeMIS structure. a) FeMIS geometry consisting of the Si substrate (1) SiO$_2$ interface buffer layer (2), bottom interface layer (3), Si:HfO$_2$ layer (4), top interface layer (5), top interface (6) and TiN top electrode (7). b) TCAD device mesh with densely meshed regions at interfaces.

6 FIELD CYCLING OF THE FERROELECTRIC HIGH-K MATERIALS

A triangular excitation voltage sweep was applied to simulate the current-voltage and corresponding polarization-voltage (integral of the transient current density) characteristics. Beside the ferroelectric, the resulting response of a purely dielectric layer was included for comparison (Figure 6.22c). Introducing the semiconductor (MFS structure) resulted in an asymmetric I-V characteristic as well as in an imprint (shift along the voltage axis) of the ferroelectric due to the internal bias field induced by the different workfunctions. Further, the introduction of the interface buffer layer (MFIS structure) created a voltage divider, which decreases the total field seen by the ferroelectric layer. The resulting sub-loop (non-saturating) operation of the ferroelectric, decreases P_r and the memory window. Directly from this simulations and transition from an MFM to an MFIS device, the influence of the voltage divider can be seen. Thus, urging the need for the application of the higher voltages. The parameters used in the simulation are given in Table 6.3.

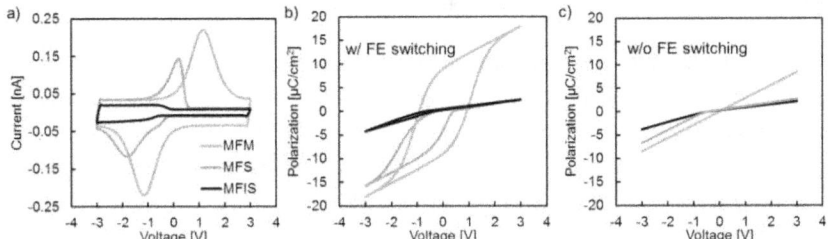

Figure 6.22 From MFM towards a MFIS/transistor. Simulated a) I-V and b) P-V characteristics of MFM, MFS and MFIS device. c) P-V characteristics of MFM, MFS and MFIS device without ferroelectric switching.

Table 6.3 Parameters used for TCAD simulations of the ferroelectric behavior within MFM to MFIS stack.

layer	d [nm]	k	P_r [µC/cm²]	P_s [µC/cm²]	E_c [MV/cm]
HfO_2	10	30	10	10.1	1
SiO_2 or $SiON$	1.2	3.9	/	/	/

In contrast to previous simulations, the FeFET was simulated as a 2D structure (see Figure 6.23) to decrease computation time. In addition to the FeMIS structures, highly doped n+ implants were defined as source and drain regions and connected to metal electrodes (see Figure 6.23).

6 FIELD CYCLING OF THE FERROELECTRIC HIGH-K MATERIALS

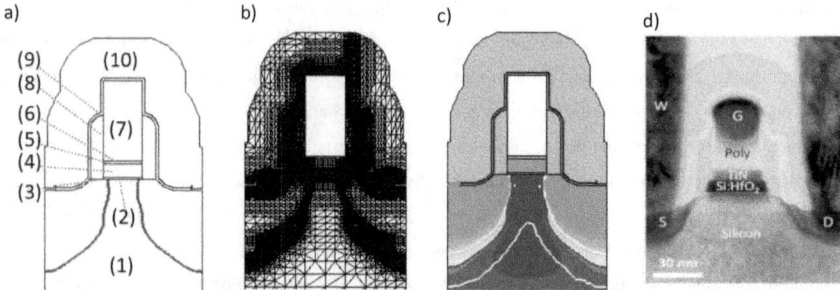

Figure 6.23 2D geometry of the multi grain FeFET device. a) FeFET geometry consisting of the Si substrate (1), SiO$_2$ interface buffer layer (2), bottom interface layer (3), Si:HfO$_2$ layer (4), top interface layer (5), TiN metal gate (6), poly Si (7), SiN spacer (8), SiO$_2$ liner (9) and nitride layer (10). b) TCAD device mesh with densely meshed regions at interfaces. (c) Geometry of the developed 2D grain boundary model of the FeFET structure. d) TEM micrograph of integrated FeFET device in 28 nm technology node.

At the end, a simulation study was performed on the FeFET structure (see Figure 6.24a) using different remnant polarization values for the ferroelectric. Figure 6.24a shows the P_r influence on the ferroelectric behavior and consequent memory window evolution which verified the behavior observed in Figure 6.17. Analog to the FeCap, the increase of P_r caused by a partial transformation of non-switching grains into switching O-phase yields an opening of the memory window.

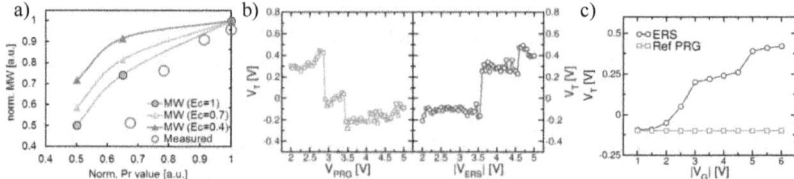

6.24 a) Comparison of measured and simulated memory window as function of remnant polarization. Simulated P_r and MW are normalized with respect to the maximum simulated P_r and MW, respectively. Simulation was performed for the P_r values of 2, 4 and 10 µC/cm². Measured Pr and MW are normalized with respect to the maximum measured P_r and MW, respectively. b) Discrete switching events for both PRG and ERS of the cell. c) Simulation of the discrete switching based on the 2D multi-grain TCAD model of FeFET. Figures b) and c) taken from [169].

In the next step, a step-wise switching was simulated. Mulaosmanovic *et al.* [169] reported a single-domain switching in ultra-scaled Si:HfO$_2$ based FeFET which results in a stepwise change in the V_{th}-characteristics shown in Figure 6.24b, for both program and erase pulses. The author performed the measurement as following: after initial 100 bipolar cycles used for device preconditioning and reach of the fully woken-up state, a reference ERS pulse (defining the high reference V_{th}) was applied; after the read-out of the reference

6 FIELD CYCLING OF THE FERROELECTRIC HIGH-K MATERIALS

V_{th}, PRG pulse was incrementally increased (50 mV steps) to study the switching to the low V_{th} state. It should be noted that every ERS and PRG operation was followed by a read-out. Vice versa sequence was applied for the ERS sweep (blue characteristics of the Figure 6.24b). With an average grain diameter in the order of 30 nm, these ultra-scaled devices (gate length = 28 nm and width = 80 nm) are expected to contain just two to four grains. Each step in the characteristics corresponds to a discrete switching event which abruptly shifts the V_{th}. To verify the model and physical process behind discreet switching events, a simulation was performed. A three-grain device was utilized for the simulation of the discrete switching. Each grain possessed a different E_c value (1 MV/cm; 1.5 MV/cm and 2 MV/cm). In Figure 6.24c, the simulation results for the 2D multi-grain TCAD model are given. It can be seen that similar to the measured characteristics, steps in the V_{th}-V_G dependence are visible. In contrast to the measured abrupt switching steps, simulated device resulted in rather smeared out switching steps. This is a consequence of the Preisach based model provided in TCAD as discussed in Section 2.3. It should be noted that in contrast to Landau-Ginzburg-Devonshire model characterized with abrupt switching, the Preisach-model which sums up discrete square-shaped hysterons and averages it over the parameter distribution (e.g. internal bias fields of coercive fields). Thus, even within the hystereses of the three single grains, sub-loops and partial switching are allowed, which results gradual shift of the V_{th}. It can be debated if this is physically justified and how small a single hysteron is. It could represent a single unit cell or also a single grain. The results found here, suggest that the latter assumption is more appropriate. Using a Preisach-model that accounts for a large ensemble of domains is no longer suitable. Instead, it can be concluded that for the purposes of the simulation of the discrete events and addressing of the single domain switching kinetics a Landau-Ginzburg-Devonshire or Landau-Khalatnikov (see Section 2.3 for details) model of ferroelectrics is necessary.

Within this section the determining parameters of the memory window were investigated. The performed simulations confirmed the strong influence of the remnant polarization on the memory window just like in the FeCap case. Moreover, a transition study from FeCap towards FeFET was performed. Replacing the metal bottom electrode by a semiconductor and adding an interface buffer layer gave rise to a voltage divider reducing the field across the ferroelectric and introduced a depolarization field. Both effects lead to a reduced P_r compared to the FeCap case. Therefore, increase of the operation voltages is needed until saturation polarization is achieved. However, further increase of, already high field over the stack would additionally accelerate the generation of the defects; result in a direct tunneling to the conduction band and thus, decrease the MW additionally. Besides, it should be noted that the breakdown field of the SiO_2 interface buffer layer (IBL) as well as the k-value ratio FE to IBL represent the upper constrain for the maximum possible field across the ferroelectrics.

6.5 Summary and Outlook

Within this chapter, the mechanisms responsible for the field cycling behavior of HfO_2 based ferroelectric memories were studied. Two main stages of the devices lifetime, namely wake-up and fatigue, were investigated. In Section 6.1., the wake-up effects were studied through comprehensive electrical characterization of internal bias fields, which were shown to be responsible for the two switching peaks observed for the pristine device. A combination of comprehensive ferroelectric switching current experiments at elevated temperatures, Preisach density analysis (FORC plots), trap defect spectroscopy and modeling of the leakage current were used to identify that no defects are generated but rather already existing defects are redistributed within the device in this initial stage of lifetime. The faster diminishing of the internal bias field at elevated temperatures solidified the assumption that the removal of the internal field, i.e. the creation of the uniform field distribution, during the wake-up was diffusion- and drift-driven. In addition, a comprehensive TEM study suggested a phase transition of the interface regions during cycling. Moreover, it also revealed the phase transformation in the bulk from predominantly M-phase in the pristine case to O-phase in the woken-up case. These results motivated comprehensive modeling of the vacancies and ion movement as well as of their recombination within the stack. By combining the modeling of the diffusion mechanisms and phase transition, it was shown that both are responsible for:

1) the disappearance of the internal bias field,

2) the creation of the more uniform field within the device stack and

3) the consequent increase of the volume fraction taking part in switching process. Modeling of the ferroelectric behavior by means of TCAD Sentaurus Device matched the measured ferroelectric behavior. Furthermore, the root cause of the limited endurance of the doped HfO_2-based ferroelectric MIMs was identified. The dielectric degradation reduces the ferroelectric switching. The increase of the defect density with field cycling indicates that the main mechanism responsible for the degradation of the ferroelectric behavior besides domain pinning is the defect generation. This degradation occurs within the interfacial regions close to the electrodes as well as within the bulk. Electron trapping at these defects affects the field distribution within the stack reducing the field in the bulk of the ferroelectric layer. Further generation of vacancies creates leakage paths, finally resulting in a breakdown of the stack before the memory window closes completely.

Even though the initial increase of polarization during the wake-up stage could be omitted by utilization of oxide electrodes and improvement of the interface properties, the issue of the limited endurance remains. In contrast to the perovskite-based ferroelectrics, ferroelectrics based on binary oxides are characterized by a very high E_c (about 30 times higher with respect to the PZT based FE [180,181]). The FeCap as the storage element used in the FeRAM applications struggles the most with the high-fields needed to overcome the

6 FIELD CYCLING OF THE FERROELECTRIC HIGH-K MATERIALS

energy barrier altering the polarization state, which is coupled to the high E_c, and still ensures endurances of 10^{15} and more [182]. Consequently, one should directly conclude that reduction of the E_c would result in decrease of the operation voltages and thus decrease of the stress experienced by the device.

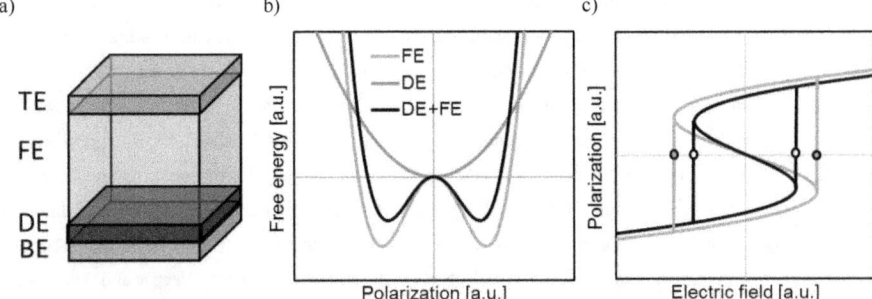

Figure 6.25 Decreasing of the barrier/E_c by utilizing the dielectric buffer capacitor (DE): a) Proposed dielectric stack of the FeRAM comprising TE/FE/DE/BE where DE represents the linear capacitor. b) Free energy vs. polarization for the ferroelectric (orange), linear dielectric (blue) and combination of both (black). c) Corresponding P-V characteristics of pure FE (orange) and combination of FE and DE stack (black).

To address this issue three paths are possible:

a) Ideally, the introduction of the suitable dopant together with usage of the inert electrodes would be the first option. Following the techniques for improvement of the intrinsic properties of the dielectric discussed in Chapter 5, introduction of the inert, noble metal electrode would strongly reduce the pullout of the oxygen from the film and increase the reliability of the film (higher VBDs and overall lower leakage). In addition to the noble metal electrode, incorporation of the new, CMOS compatible dopant or even a material mixture which would result in a closer to ideal (squared shaped) P-V characteristics with lower E_c (below 1 MV/cm), would reduce the operating conditions and drastically increase the endurance of the device.

b) Second option is modification of the stack by addition of the interlayers. As discussed before retention and access time (governed by applied field) are inversely proportional parameters. Hence, that potential decrease of the E_c would result in a decrease of the voltages needed to set desired polarization state, but at the same time decrease of the retention of the device. Müller et al. [137] reported retention of the hafnia based FeCaps which easily fulfills the targeted 10-year margin. Therefore, small "sacrificing" of the retention by modification of the stack via addition of the linear capacitor under and/or above the ferroelectric film, thus forming the device consisting of the TiN TE/Sr:HfO$_2$/HfO$_2$/TiN BE would be a second approach. This approach is depicted in Figure 6.25a. Utilizing this approach a parabolic potential energy of the pure paraeletric layer/interface modifies the double well potential of the ferroelectric (e.g. Sr:HfO$_2$) as shown in Figure 6.25b). Combination of these two materials results in a lower effective barrier seen by the ion and consequent reduction of the E_c. Lower E_c enables usage of the lower voltages for altering the state of the FE. On the other side, additional paraelectric layer behaves like a dead layer/resistance in series to the FE layer. Hence, a dead layer is causing additional voltage drop and might cause usage of the higher operation fields. It can be concluded that the combination of these two stack constituents should be chosen with high precision in order not to end up with two diametric outcomes, a decrease of the E_c and increase of the operation condition or voltage needed for the switching of the ferroelectric.

c) The third possibility would be usage of the modified anti-ferroelectric materials that are characterized with higher breakdown strength and much higher endurances [74]. In addition, the critical field needed for the induction of the FE phase within nominally unipolar films can be rather low with respect to the E_c of the ferroelectric. On the other side, anti-ferroelectric and field-induced ferroelectric materials cannot be directly utilized as memory films due to the already mentioned loss of the state upon removal of the external field. This problem can be overcome, as will be shown in following Chapter 7 where the novel concept of non-volatile anti-ferroelectric memory will be presented.

7 Anti-ferroelectric Non-volatile Memory

Reliability and stable field cycling behavior are among the most important requirements of modern non-volatile memories. In the previous Chapter 6, the utilization of doped HfO_2 ferroelectric for emerging nonvolatile memories was discussed. It was shown that changes of memory window size with field cycling and limited lifetime can be correlated with:

a) local internal bias fields
b) oxygen vacancy generation and re-distribution driven phase transition
c) ion displacement in ferroelectric memories that induces additional stress in the device's crystal lattice, which results in a decrease of the device lifetime.

The very high electric field needed to overcome E_c ($\approx 1 - 2$ MV/cm) creates defects which facilitates hard breakdown. In addition, polarization switching events cause an additional field increase to the already high fields in the material stack. Binary oxide ferroelectrics such as doped HfO_2 have a typical breakdown voltage that is just about two times higher than their coercive voltage. With regard to the endurance, two possible improvements can be considered: a material with either higher breakdown field strength or lower E_c is needed. Bearing in mind that anti-ferroelectric materials exhibit longer lifetimes and better field cycling stability (see Figure 2.13; Chapter 2.4), in this chapter anti-ferroelectric like materials are proposed as the basis of a new memory concept with extended endurance. Further, a detailed material study moving from material towards a new device concept is given. It should be kept in mind that, in contrast to ferroelectric materials with non-zero P_r, AFE films such as 10 nm thick ZrO_2 cannot be directly used for non-volatile data storage, because their P_r is ≈ 0. In other words, reducing the external field to zero causes a depolarization, resulting in a complete loss of the stored information.

To address these fundamental challenges a new device concept based on an asymmetric stack will be presented. Here, the internal bias field, one of the nemeses of the HfO_2 based ferroelectric discussed in Chapter 6, will be employed to center one of the switching branches of the anti-ferroelectric double-hysteresis on zero bias. Further, the introduction of an internal bias field is shown to make the P_r of an anti-ferroelectric material non-zero, thus enabling non-volatile data storage. Furthermore, based on the Landau-Ginzburg-Devonshire theory and an experimental proof of concept, this novel memory concept is discussed. Finally, reliability characterization of this proof of concept based on asymmetric electrodes will be presented.

Certain parts of the text and results given within this chapter contain the findings presented after the peer-review process in [61], and [77].

7.1 On how Classic DRAM Material ended up Anti-ferroelectric

The goal of endurance extension of polarization memories should be based on the materialization of the layer quality and scalability offered by DRAM stacks on one side and fatigue strength characteristic for the anti-ferroelectric like materials. As discussed within the introduction (Chapter 1), DRAM stacks are well known for their exceptionally high durability and very high k-value [183]. This high k-value is common characteristic of materials in which the centrosymmetric cubic phase is stabilized [183]. In addition to the cubic phase, the tetragonal phase characterized as having the second highest k-value is also reported in state-of-the-art DRAM ZrO_2 stacks [61]. In contrast to the centro-symmetric cubic phase, which does not support any kind of the ion displacement without applied external electric fields i.e. spontaneous polarization, it is reported in refs. [69] and [184] that the tetragonal phase is responsible for the anti-ferroelectric behavior within ZrO_2. This, together with the study that reported non-interruption of the ferroelectric properties by insertion of the interlayer [185] motivated the broad range[20] C-V and polarization measurements on the ZAZ capacitors. Indeed, during the small-signal capacitance measurements of TiN/ZAZ/TiN (classic DRAM devices described in Chapter 3.1) capacitors an asymmetric butterfly-like C-V curve, characteristic for anti-ferroelectrics, was observed (see Figure 7.1a). Nonlinear k-value variation as a function of the applied DC field as well as the hysteretic behavior, shown in Figure 7.1a, indicates the presence of some portions of the tetragonal phase which, as reported by Materlik et al., is responsible for exhibiting field-induced ferroelectric behavior in thin ZrO_2 films [69]. Reasonably, wherever the butterfly C-V curve is present, it can be expected that the polarization response to a triangular voltage excitation should exhibit hysteretic behavior. In Figure 7.1b I-V and P-V hysteresis curves obtained from triangular field sweeps are presented. Looking at the I-V characteristics, a strong asymmetry of switching and backswitching current peaks on negative (A and A') and positive (B and B') polarity is observed (Figure 7.1b). Both, the asymmetry of the switching/backswitching current peaks as well as the shift of the pinched hysteresis curve can be attributed to the asymmetric dielectric stack where the Al_2O_3 interlayer is located closer to the top electrode. The different thicknesses of the sandwiching ZrO_2 layers result in different crystallinities and therefore, different fractions of the tetragonal phase in both layers.

---

[20] Typically, C-V measurements on the DRAM structures are performed with amplitudes below 1.5 V. Here, a broader sweep of 3V was applied.

7 ANTI-FERROELECTRIC NON-VOLATILE MEMORY

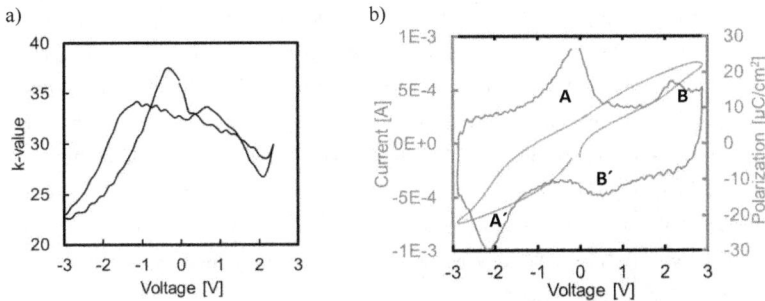

Figure 7.1 a) *k*-value-voltage characteristics exhibiting strong nonlinear hysteretic behavior characteristic for anti-ferroelectric materials. b) Current-voltage and corresponding polarization-voltage characteristic of a TiN/ZrO$_2$/Al$_2$O$_3$/ZrO$_2$/TiN DRAM capacitor. A/A' and B/B' denote a switching/backswitching pairs occurring on negative and positive polarity, respectively.

In order to assess the structural properties of the ZrO$_2$/Al$_2$O$_3$/ZrO$_2$ film, confirm the assumption that the tetragonal phase is predominantly responsible for the observed switching and AFE behavior and engineer its properties, a material study will be presented in following chapter.

7.2 Stabilization of the Tetragonal Phase in ZrO$_2$ Thin Films

As a first step to identify the film morphology and resulting AFE behavior, the complexity of the examined stack has to be reduced. Therefore, a detailed material study was performed on samples with a simpler structure. In search for optimal condition (see section 7.4.2 for details), ALD deposited MIM capacitors comprised of a single ZrO$_2$ layer sandwiched between two metal electrodes (see detailed description in section 3.1) were crystallized in N$_2$ atmosphere at three different annealing conditions (450 °C[21] for 2 minutes; 500 °C for 2 minutes and 800 °C for 20 seconds). It should be noted that the annealing conditions were chosen to be compatible with the thermal budget of classic DRAM capacitor (450°C for 10 minutes [183]), processed in the back end of line. As it can be seen in Figure 7.2, a lower[22] thermal budget resulted in stronger FFE behavior compared to the higher one. Larger AFE switching peaks are observed as well as a higher ratio of the switching current with respect to the pure dielectric response[23].

---

[21] First annealing condition (450°C for 2 minutes) yielded a paraelectric behavior. However, this anneal condition was disregarded from the further analysis due to uncertainties brought by the unstable temperature control, commonly observed in NaMLab, rapid thermal processing unit when utilized for a long anneals, below 500 °C.
[22] Within the scope of this thesis thermal budged is defined as $\sqrt{Anneal\ temperature\ [°C] * Anneal\ time\ [s]}$
[23] At this point it can be concluded that increase of the annealing temperature has a stronger influence on desired phase stabilization in contrast to the duration of the annealing step. This assessment will be confirmed latter in section 7.4.2 when additional anneal at 650° C was performed.

In addition to ALD-deposited films, ZrO$_2$ based capacitors were grown with PVD. The sputter deposited ZrO$_2$ based capacitors were grown under different Ar pressure conditions. XRD patterns given in Figure 7.2e show that ZrO$_2$ films deposited under lower Ar pressure conditions resulted in amorphous phase with a slight monoclinic component whereas films at higher Ar pressure were amorphous. Especially, the peaks located at a 2Θ angle of 28° confirm this assessment. A deposition at lower Ar chamber pressure results in a higher kinetic energy of the Ar ions in the plasma relative to a higher energy of the sputtered particles. This results in partial crystallization during the PVD process and an increase of the monoclinic phase portion in the film. After capping the stack with the top electrode, a crystallization step via rapid thermal annealing was performed.

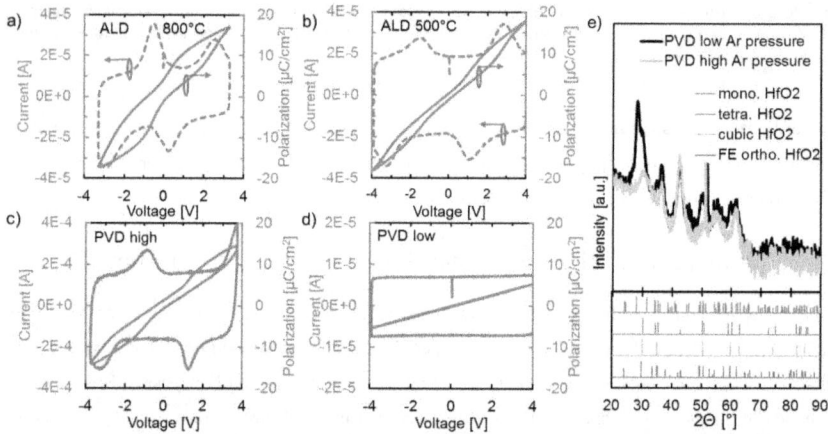

Figure 7.2 Electrical and physical properties of the ALD and PVD deposited TiN/ZrO$_2$/TiN capacitors. P-V and I-V characteristics of ALD deposited films annealed at a) 800 °C for 20 s and b) 500 °C for 2 minutes in N$_2$ atmosphere. P-V and I-V characteristics of PVD grown ZrO$_2$ films, deposited at c) high Ar and d) low Ar pressure. e) XRD patterns of the as deposited PVD thin films at high (gray line) and low (black line) Ar pressure. Figure taken from study by Pešić et al. [61].

To assess the electrical response and AFE properties of the PVD deposited films, P-V measurements were performed. In Figure 7.2c-d, I-V and corresponding P-V comparison of TiN/ZrO$_2$/TiN structures deposited with high and low Ar pressure is given. The MIM capacitor deposited under lower pressure conditions exhibits pure paraelectric (Figure 7.2d) behavior. This device exhibits only a dielectric response without switching current peaks which translates into linear paraelectric P-V characteristics. In contrast to the low pressure deposited stack, the MIM stack deposited under high pressure is characterized with double current peak characteristics (as well as pinched hysteresis) as known for field-induced-ferroelectric materials.

7 ANTI-FERROELECTRIC NON-VOLATILE MEMORY

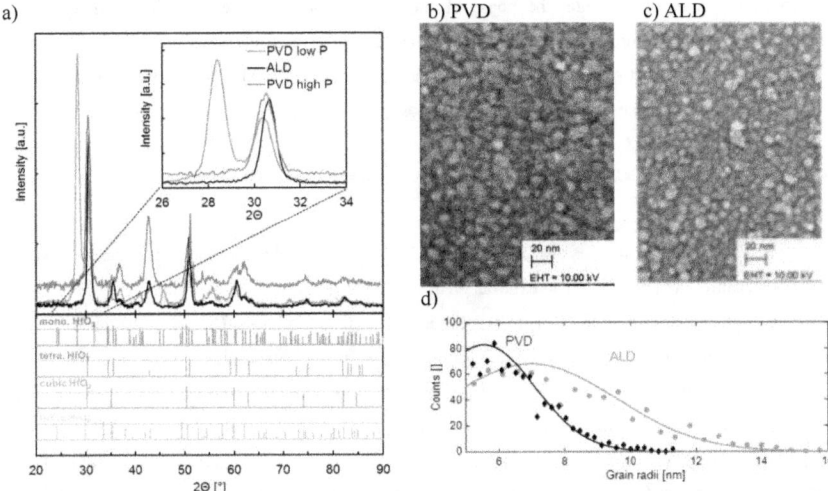

Figure 7.3. a) X-ray diffraction patterns of ALD deposited and sputter-deposited ZrO_2 films compared to the peak positions of the monoclinic and tetragonal phases in ZrO_2. b) SEM top view of 10 nm thick sputter deposited (PVD) and c) 12 nm thick ALD deposited ZrO_2 thin films. d) Distribution of grain radii extracted from SEM measurement (symbols) of b) and c) with Gaussian fit of grain radii (solid lines). Figure taken from study by Pešić et al. [61].

The film morphology and the structural properties of the two ZrO_2 layers fabricated by ALD and PVD were assessed by means of XRD and scanning electron microscopy. XRD measurements were performed and compared to the patterns for the monoclinic, orthorhombic, cubic and tetragonal phase patterns of ZrO_2 (Figure 7.3a lower part). ALD deposited ZrO_2 films showed a tetragonal phase pattern that was previously reported as responsible for the AFE behavior [69]. Similarly, films deposited by means of PVD under high pressure also exhibited stabilization of the tetragonal phase. In contrast to those two, the stack that was partially crystalline after the deposition (PVD at low Ar pressure) resulted in a mixture of the previously stabilized monoclinic and post deposition annealing induced tetragonal phase. A zoom into the most dominating peaks at a 2Θ angle of 30° confirms this assessment. Therefore, it can be concluded that similar film morphology for both PVD[24] and ALD grown films is responsible for the stabilization of the tetragonal phase and AFE P-E behavior characterized by pinched-hysteresis loops. SEM scans were performed after the removal of the top electrode on the bare oxide (see Figures 7b-c). The grain radii were determined from SEM top view images using the software *Gwyddion* (Figure 3b) [186]. The Gaussian fit of the data shows a in similar average grain radii of 7±3 nm for ALD and 5.7±2 nm for PVD deposited films. As simulation

---

[24] Sputter deposited (PVD) film at high Ar pressure.

7 ANTI-FERROELECTRIC NON-VOLATILE MEMORY

results by Materlik et al. indicate, the phase of the material and by extension the ferroelectric/anti-ferroelectric behavior, is mainly determined by the surface energy effects related to grain sizes [69]. Having resolved the structural questions and the phase transition responsible for AFE behavior in ZrO₂ thin films, possible applications of these properties should be assessed.

7.3 Intrinsic Properties of Anti-ferroelectric Materials

As mentioned above, the hysteresis curve both of standard ZAZ DRAM and of pure ZrO_2-based capacitors displays polarization that depend on an external electric field (e.g., through an external voltage) and exhibit a "pinched" hysteresis loop (Figure 7.2a or Figure 2.6c). The corresponding P-E characteristic has a positive polarization hysteresis in the positive voltage regime and a negative polarization hysteresis in the negative voltage regime. Furthermore, the charge is linearly related to the voltage in the vicinity of zero field. The two theories explaining the fundamental physical mechanisms governing the behavior of AFE films were discussed in Section 2.3. Independently of the theory (either antiparallel dipoles [58] or field-induced phase transformation [187,188,60] theory) the effect (polarization) exhibited is volatile, and thus, with removal of external field, an anti-ferroelectric material loses its state (remnant polarization is zero). Furthermore, this possible phase transition (exhibiting of the positive or negative branch of the pinched hysteresis) is symmetric to zero field and polarization and can be induced for both voltage polarities independently.

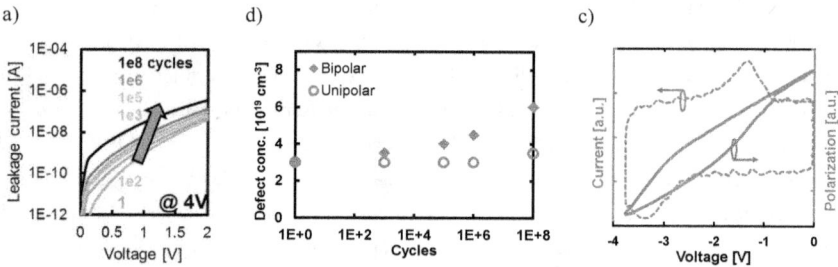

Figure 7.4. a) Field cycling (bipolar) dependent DC leakage current measurement used for defect concentration extraction. b) Defect concentration evolution as function of unipolar and bipolar cycling performed at 1 kHz. c) Unipolar current-voltage (dashed) and corresponding polarization-voltage (solid) characteristics. Figure taken from study published by Pešić et al. [61].

In section 6.3 it was shown that unipolar stress generates significantly lower degradation with respect to the bipolar stress in HfO_2 based ferroelectrics [21]. Therefore, unipolar switching and backswitching as the unique property of anti-ferroelectrics should be considered for potential memory devices. As in the case of doped HfO_2, a bipolar stress is compared to the unipolar stress by means of leakage current trap spectroscopy (Section 4.1) to extract the defect concentration evolution with field cycling. Comparing the defect density evolution with unipolar vs. bipolar cycling in Figure 7.4b, it can be seen that the onset of the degradation in the unipolar case is found at much higher number of stress cycles with respect to the case of bipolar stress. Accordingly, reduced film degradation under unipolar switching conditions would be preferred in device applications. Unipolar switching/backswitching between the polarized and unpolarized state can be seen in the Figure 7.4. The resulting polarization hysteresis is similar to FE switching between the positive and negative polarized state, but is shifted along the voltage axis and observable either on negative or positive polarities.

In contrast to FE materials with non-zero remnant polarization P_r, ZrO_2 based materials cannot be directly used for non-volatile data storage. Reducing the external field to zero causes depolarization, resulting in a loss of the stored information. To address these fundamental challenges, a new concept that engineers an asymmetric stack to introduce an internal bias field, that centers one of the branches of the pinched hysteresis characteristic is presented. This approach is motivated by the hysteresis field shift observed in Sr:HfO$_2$ ferroelectrics caused by asymmetric electrodes which was discussed in Chapter 5.3. In other words, utilization of the very internal bias field, which is the nemesis of ferroelectric materials, produces a non-zero P_r in an AFE material and enables non-volatile data storage. In the following, a theoretical analysis, modeling and realization of the world's first anti-ferroelectric non-volatile memory is presented.

7.4 Building a Anti-ferroelectric Non-volatile Memory

Since there is no remnant polarization at zero bias, the anti-ferroelectric hysteresis cannot be used for non-volatile memory applications (see Figure 7.4c). In order to address this fundamental challenge a theoretical analysis was performed. Afterwards the development of an anti-ferroelectric based non-volatile memory cell is presented. Finally, based on the theoretical analysis and simulation results first anti-ferroelectric non-volatile memory (AFE-RAM) devices were fabricated and characterized.

7.4.1 Theoretical Analysis and Modeling of AFE-RAM

The theoretical background describing the introduction of the internal bias field and its consequences on the free energy were addressed by Pešić et al. [61]. The energy potential landscape described by LGD formalism in Section 2.3 is revisited and extended. According to the LGD equation, a ferroelectric double well potential and an anti-ferroelectric single well potential look as shown in Figure 7.5a and Figure 7.5b, respectively if the external field E and thus, the whole term $-E \cdot P$ in equation 2.11 equals zero. Precisely this term is of central interest in the following, as it introduces the electric field dependence into the equation governing the model. This external electric field represents the effective electric field experienced by the ferroelectric or anti-ferroelectric material. Remembering the discussions of Sections 5.1 and 5.3, this effective field consists of the applied field at the electrodes and the internal bias field due to the workfunction difference of the electrodes. Thus, equation 2.11, expressing Free energy Φ modifies to:

$$\Phi = \frac{1}{2}\alpha P^2 + \frac{1}{4}\beta P^4 + \frac{1}{6}\gamma P^6 - E_{external}P - E_{built\_in}P = \Phi_{AFE} - E_{built\_in}P \quad (7.1)$$

, where P represents the ferroelectric polarization, (α,β,γ) the Landau expansion coefficients, $E_{external}$ the applied electric field, E_{built_in} the internal bias field and Φ_{AFE} the free energy of anti-ferroelectric without the internal bias field.

As a consequence of this internal bias field, the single-well potential of the AFE material can be modified to a double well potential by introducing an energy barrier (see Figure 7.5a). Depending on the initial potential shape of the AFE material, this barrier between the wells can be high enough to separate the polarized/non-polarized state similar to the barrier separating positive and negative polarization state in a FE material. This situation centers one of the two AFE *P-E*-loops around zero field favoring one of the polarization states. This occurs at the expense of the other polarization state, whose corresponding loop moves further away from zero field toward fields so high that they are not of interest in the following discussion. By applying an external field, an AFE material with an internal bias as described can be switched similar to an FE material. However, the states that the material is switched between are not the positive and the negative polarization state as in the case of an FE material (see Figure 7.5a) but, instead, it

7 ANTI-FERROELECTRIC NON-VOLATILE MEMORY

is switched between the positive (or negative, depending of the sign of $E_{built-in}$) and the non-polarized state (Figure 7.5b-d). This situation is anticipated to drastically improve the device lifetime and endurance[75,74]. The possible causes for improved cycling behavior of AFE materials compared to FE materials will be discussed in detail in section 7.5.1.

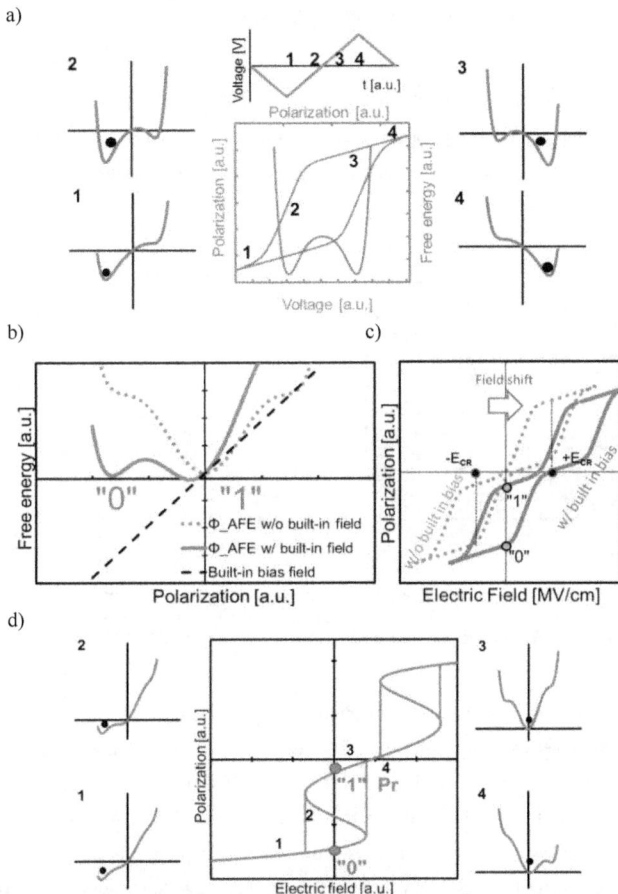

Figure 7.5 Free energy (solid) of a) a ferroelectric material as a function of the applied field and resulting hysteresis curve. b) Free energy and corresponding c) polarization hysteresis of an anti-ferroelectric material at zero external fields with (solid) and without (dotted) internal bias filed. The shift in the polarization characteristics along the voltage axis results in a remnant polarization and therefore enables non-volatility. d) Free energy of an anti-ferroelectric material with internal bias field as a function of the applied external field together with the corresponding non-volatile AFE like hysteresis (center). Figure taken from study published by Pešić et al. [61].

7 ANTI-FERROELECTRIC NON-VOLATILE MEMORY

Beyond the intentional choice of electrodes with different workfunctions, an internal bias field can also be generated by either the insertion of a fixed charge in an asymmetric position with respect to the middle of the anti-ferroelectric thin film or by introducing an electrical dipole layer [189] at an arbitrary position of the layer. Even though this is not the main focus here, it should be kept in mind especially in the light of the role of defect generation and movement established in Sections 6.1 and 6.2.

The suggested approaches are suitable for creation of an internal (built-in) bias field that essentially shifts the polarization-voltage characteristic of the anti- or field-induced-ferroelectric material along the voltage axis. Utilization of a high workfunction top or bottom electrode (e.g. Pt ~ 5.6 eV) [175] while maintaining a lower workfunction electrode on the other side would yield in the required internal bias voltage and thus result in the desired centering of the hysteresis.

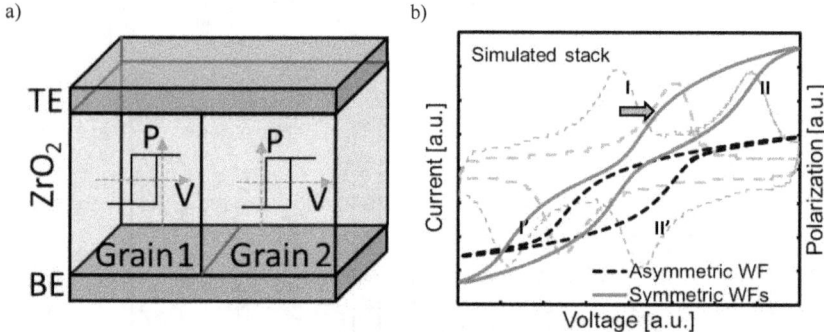

Figure 7.6. Modeling the capacitor of the AFE-RAM. a) Illustration of the 3D device Landau-Khalatinikov compact model utilized for the simulation of the AFE behavior. b) P-V and I-V characteristics of the simulated structure with symmetric (solid) and asymmetric (dashed) material stacks. Figure taken from study by Pešić et al. [77].

Before practical realization, a theoretical verification of the suggested approach is needed. Therefore, a compact multi-grain model of the AFE-RAM cell was developed in MATLAB and TCAD. Modeling of the anti-ferroelectric behavior is achieved in the following way: Two oppositely biased FE grains reproduce the AFE I-V characteristics and corresponding P-V hysteresis (Figure 7.6). This is similar to the scenario shown in Figure 7.6a. Indeed, this situation not correctly reflect what happens at atomic scale, but it results in the same macroscopic electric response which is all that is needed in the simple model. By exchanging the TiN top electrode with a material with higher WF (about 1 eV higher), a field shift of the whole I-V characteristic is induced. As explained before, the switching/backswitching pair I/I' is centered around zero, whereas the second switching/backswitching pair II/II' is shifted to fields beyond the intended operation regime of the device (see Figure 7.6b). Consequently, a centering of one of the hysteresis loops is achieved.

When switching the device between zero and negative polarization, its behavior is indistinguishable from an FE material exhibiting just half the spontaneous polarization.

7.4.2 Practical Realization of the AFE-RAM

Based on the theoretical analysis and simulation results, the fabrication of the world's first AFE-RAM device and its characterization are presented in the following. Before deposition of the top electrode and consequent introduction of the internal bias, the AFE properties of the material itself have to be addressed. Several design strategies have to be considered. For low power operation, the occurrence of the current peaks should not be characterized with critical/coercive fields that are too high. Since the AFE-RAM cell will be operated in same manner as FeRAM, in the reminder of the text, the critical field E_{cr} needed for the phase transition will be addressed as E_c. Beside the low power operation, a lower E_c is directly correlated with less stress and defect generation since device operation voltages are reduced. Furthermore, the peaks i.e. I /I' (Figure 7.7), should be in the vicinity of the zero fields, so that the i.e. positive (TE, higher WF value), workfunction difference can generate a sufficient bias field needed for the centering of the half hysteresis branch. In addition, the workfunction difference between the top and bottom electrode should not be too high, since it would negatively impact the charge injection into the film. Asymmetric injection results in imprint (as discussed in Section 6.4.1) and in general accelerates the fatigue of the device. Thus, it can be seen that not all AFE characteristics are favorable for the potential AFE-RAM. To address these design questions and tailor the material properties, an annealing study was performed. In addition to stacks discussed in section 7.2 (stacks annealed at 500 °C for 2 minutes and 800 °C for 20 s) a slightly lower thermal budget is tested (sample annealed at 650 °C for 20 s) after which the P-V characteristics were measured and correlated with SEM measurements. The focus was on the field occurrence of the peaks (field needed for reaching the certain switching/backswiching pair) and on the distance between the switching (V_I) and backswitching ($V_{I'}$) current peak (see Figure 7.7b), i.e. I and I', respectively. Interestingly, an increase in the annealing temperature yielded a decrease in the grain size. Moreover, in Figure 7.7e can be seen that with the increase of the annealing temperature, the distance between the positive and negative switching/backswitching current peak pairs decreased. Consequently, an increased annealing temperature resulted in a lower critical/coercive field at which the field-induced (tetragonal to orthorhombic) phase transition occurs. Comparison of thermal budged reveals that the lowest thermal budget has a has better/worse AFE characteristics in contrast to highest/middle thermal budget. This once again confirms assessment that the annealing temperature and not the time is crucial for stabilization of AFE behavior.

7 ANTI-FERROELECTRIC NON-VOLATILE MEMORY

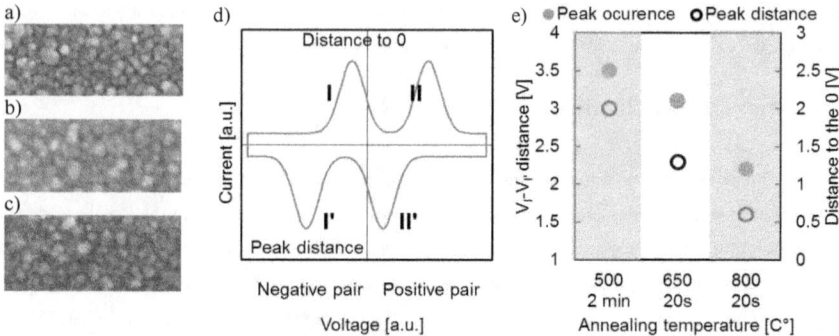

Figure 7.7 a) SEM top view images for ZrO₂ based capacitors annealed at a) 500 °C for 2 minutes b) 650 °C for 20 seconds and c) 800 °C for 20 seconds. AFE peak position d) schematic and e) dependence on the annealing temperature.

According to the discussed design guidelines and the results of the material study, ZrO₂ based stacks annealed at 800 °C for 20 seconds were selected as the best candidate. These films exhibited the most prominent AFE behavior and switching peaks in the vicinity of zero field. In the next step, as top electrode material, ruthenium oxide (RuO$_x$) was considered. RuO$_x$ is a conductive oxide and the electrode material characterized by a very high workfunction of 5.3 eV [61,190] that is about 0.8 eV higher than the workfunction of the TiN BE (TiN$_{WF}$ ≈ 4.5 eV as measured in section 5.2). The workfunction asymmetry introduced into the 10 nm stack resulted in a strong internal bias field on the order of 1 MV/cm. The RuO$_x$ TE was deposited by PVD on top of the previously discussed TiN/ZrO₂ capacitor stack. The influence of the introduction of the internal bias field by exchange of the TE can be seen in Figure 7.8. Resulting, *I-V* and *C-V* characteristics were shifted along the field axis by the internal bias field. Finally, re-adjustment of the operation condition using lower absolute voltage for the sweep yielded a centering of the hysteresis loop (see Figure 7.8b) and a *2P$_r$* value of about 17 µC/cm².

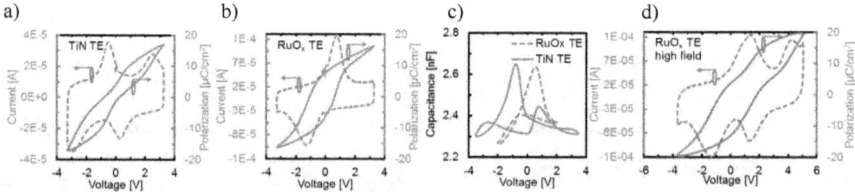

Figure 7.8. Comparison of electrical properties of TiN/ZrO₂ AFE material stacks with a) TiN top and b) RuO$_x$ top electrode: *I-V* (dashed) and *P-V* (solid) characteristics. c) Comparison of *C-V* characteristics of the TiN/ZrO₂ AFE stack with TiN (solid) and RuO$_x$ (dashed) top electrode. d) Preservation of AFE behavior within TiN/ZrO₂/RuO$_x$ stack is observed by extending the voltage amplitude to more positive fields. Figure taken from study published by Pešić *et al.* [61].

7 ANTI-FERROELECTRIC NON-VOLATILE MEMORY

The previously discussed switching between the positive/negative polarized and the unpolarized state is shown in Figure 7.8c. In contrast to ferroelectrics, which exhibit a rather symmetric butterfly C-V characteristic, here, a bigger half butterfly curve would correspond to the positive polarization state whereas the small hysteresis loop for negative voltages corresponds to the unpolarized state. It is expected that the RuO_x deposition has minimal[25] influence on the film morphology since a RuO_x TE was deposited through a shadow mask at room temperature after the tetragonal phase within the ZrO_2 film was stabilized by annealing of the previous MIM structure capped with a TiN TE. Measurement of the same TiN/ZrO_2/RuO_x stack with adjusted field sweep[26] (see Figure 7.8d) confirms the previous assumption that a comparable AFE behavior is reached for RuO_x and TiN top electrodes. Hence, it can be seen that the second pair of switching/backswitching current peaks and the corresponding second P-V hysteresis loop of the RuO_x stack are also shifted along the field axis. Moreover, utilizing the presented approach it can be concluded that no backswitching (characterized for the volatile, AFE material) occurs after removal of the external field.

In order to solidify the observed behavior, a FORC analysis was performed. Preisach/switching density of TiN/ZrO_2 with TiN and RuO_x top electrode are shown in Figure 7.9a and 7.9b respectively indicating internal bias field E_{bias} and E_c contributions for switching. Analogous to the P-V characteristics, capacitors with TiN top electrode are characterized by two switching distribution maxima that are located at different bias fields but with similar E_c (see Figure 7.9a). The utilization of the RuO_x top electrodes causes an internal bias field, which consequently shifts the switching distributions towards higher electric fields. This internal bias centers the single distribution that perfectly corresponds to the woken-up state of ferroelectric doped hafnia based capacitors. Even though the internal bias field was induced, the magnitude of coercive field was preserved.

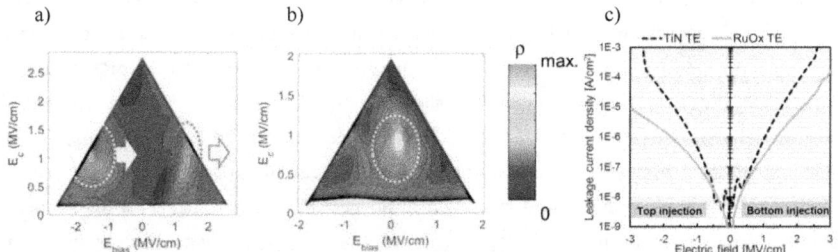

Figure 7.9 The Preisach/switching density ρ of TiN/ZrO_2/ stack with a) TiN and b) RuO_x top electrode. The circled area together with the arrows denotes the movement of the switching distribution peaks with introduction of the top electrode with a higher workfunction (RuO_x). c) Measurement of leakage current density (solid line) I-E of an ALD deposited MIM capacitor (0 to ± 3 MV/cm): TiN/ZrO_2/ stack with TiN and RuO_x Top electrode. Figure taken from study published by Pešić et al.[61].

---

[25] Even though the AFE properties are uninterrupted, a TEM study would be beneficial to investigate a possible interface formation due to the SC-1 etching step which precede the RuOx electrode deposition.
[26] By extending the amplitude of the applied voltage, a second pair of switching/backswitching peaks can be observed.

In order to investigate the influence of the internal bias field on the electrical properties, leakage currents of TiN/ZrO$_2$ stacks with TiN and RuO$_x$ top electrode were analyzed (see Figure 7.9c). These measurements were performed with a delay of 10 s between stress and measurement in order to rule out any dielectric relaxation and transient switching current effects. Analogous to the usage of Pt TE discussed in Chapter 5, increase of the WF resulted in reduction of leakage current in both injection sides. An increase of the workfunction and thus the Schottky barrier height (SBH) represents an effective way to introduce an internal bias field within the standard ZrO$_2$ based capacitor stack. Thus, the introduction of a RuO$_x$ top electrode strongly decreases the leakage current for both voltage polarities. The RuO$_x$ electrode is characterized by about 0.8 eV higher workfunction ($\approx$5.3 eV) compared to TiN ($\approx$4.5 eV) resulting in an internal bias and stack asymmetry.

7.5 Reliability and Device Performance

7.5.1 Retention and Endurance of AFE-RAM

After confirming the suggestions obtained from simulations, merits of non-volatile memories were examined. The two most important reliability characteristics of non-volatile memories are:

a) high endurance during field cycling between the program and erase state and

b) retention or ability to preserve the state after the external electric field is removed.

As discussed in Chapter 6, doped HfO$_2$ based ferroelectric memories suffer from the changing remnant polarization (memory window) with field cycling. In addition, early breakdown or closure of the memory window was observed (see section 2.5.2; Figure 2.15). In contrast to the HfO$_2$ based FeRAM case, the concept of the AFE-RAM memory shows much higher endurance. As can be seen in Figure 7.10a a stable cycling operation of 10^{10} cycles was observed. Even though an asymmetric stack was utilized, which typically results in an asymmetric injection of charges, the influence of the imprint is heavily reduced with respect to the endurance gain obtained by utilization of the AFE like potential. First, the barrier as determined by the coercive field for switching between the non-polarized and polarized state is significantly lower. A low coercive field of 0.7 MV/cm allows lower operation voltages, thus extending the lifetime of the device. In addition, the higher fatigue resistance of the AFE materials can be attributed to the different mechanism of switching compared to ferroelectrics as previously reported [74]. The stress generated by a spontaneous polarization switching in a FE is much higher than the corresponding stress within the AFE-like material due to the PE-FE phase transition. Further, it was suggested by Lou et al. [74] that a lower charge injection during switching might be responsible for less fatigue in AFE compared to FE materials.

7 ANTI-FERROELECTRIC NON-VOLATILE MEMORY

Furthermore, only half of the polarization is used, which further extends the immunity towards the field cycling. The application of an electric field that is only half of the magnitude needed for FeRAM reduces the stress on the dielectric, thus hindering the defect generation. In Figure 7.10b, a comparison of simulated band diagrams of FeRAM and AFE-RAM stacks at operating condition of 4 MV/cm and 2 MV/cm is given. Smaller band banding of AFE-RAM reduces the probability of electron injection with respect to the FeRAM device. Moreover, significantly lower operating voltages of the AFE-RAM with respect to a hafnia-based FeRAM of similar thickness are possible.

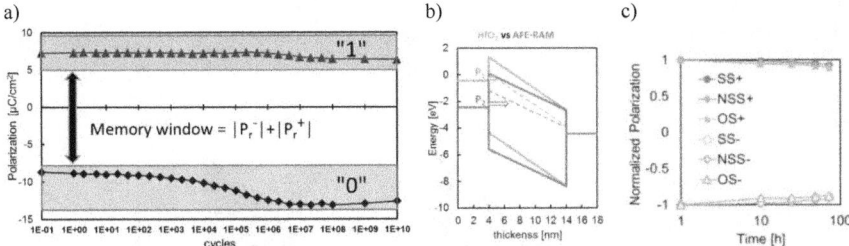

Figure 7.10 a) Endurance characteristics at room temperature. b) Band-diagram of a 10 nm FeRAM and AFE-RAM equivalents at operation condition of 4 MV/cm and 2 MV/cm, respectively. Shorter tunneling path and higher probability for charging of the FeRAM with respect to the AFE-RAM c) Retention characteristics at 100 °C. shows positive same-state (SS), opposite-state (OS) and new-same state (NSS) retention as well as their negative counterparts. Figure a) taken from study published by Pešić *et al.* [61].

In addition to the endurance characteristic, retention of this novel memory concept was examined (Figure 7.10c). According to the procedure described in Section 4.3, retention tests were performed for the samples programmed or erased with ±3 V pulses. In case of the FeRAM memories, same-state (SS), opposite-state (OS) and new-same state (NSS) retention was examined on four discrete capacitors [111]. Dedicated pulse was used to set a defined polarization state in each of these four devices, after which the devices was stored at 100 °C for different time intervals. After each bake, the subsequent sequence was applied to investigate if the state is retained (for details see Section 4.3). In Figure 7.10c it can be seen that all states are preserved for more than 75 hours, thus justifying the non-volatility of the innovative concept.

7.5.2 Field Cycling Behavior and Properties of the AFE-RAM

To gain more insight into the device properties, FORC field cycling experiments were performed. Similar, to the procedure described in section 4.2.3, the FORC measurement routine was nested into the field cycling program/erase pulse sequence. As shown in Figure 7.11a-c, the FORCs were recorded for the pristine device and after 10^9 and 10^{10} cycles. Further, the evolution of the absolute bias field of the maxima in the FORC plots of ferroelectric Sr:HfO$_2$ based FeRAM (see Chapter 6.1) was compared to the trend of the

7 ANTI-FERROELECTRIC NON-VOLATILE MEMORY

AFE-RAM in Figure 7.11d. In contrast to the double peak Preisach switching density (see Figure 5.3a) of the ferroelectric Sr:HfO$_2$ based MIM capacitor, the Preisach switching density of the AFE-RAM is characterized by a single peak in the Preisach plane, that correlates with the observed I-V characteristics. Even though the AFE-RAM exhibits a slight biasing with field cycling, it shows much higher stability with respect to the Sr:HfO$_2$ based ferroelectric capacitor. This slight biasing is due to the asymmetric electrodes and asymmetric charge injection. Besides the bias field stability, the distribution of E_{bias} and E_c in the AFE-RAM stack is much narrower compared to Sr:HfO$_2$. This narrow distribution is a valuable property needed for the improvement of the variability in polycrystalline polarization based memories (e.g. FeRAM and FeFET) since it corresponds to a more abrupt switching and consequently, sharper current peaks. Further, both device concepts exhibit stable coercive fields with field cycling.

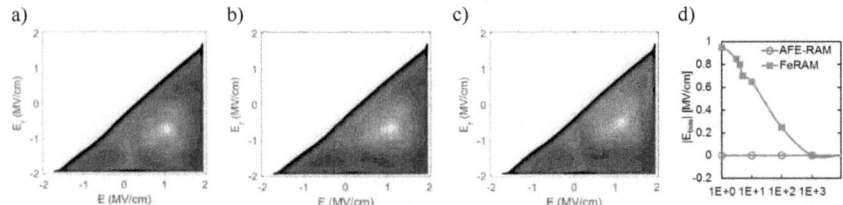

Figure 7.11 Preisach switching density plots obtained by FORC for TiN/ZrO$_2$/RuO$_x$ capacitor in a) pristine stage and after b) 10^9, and c) 10^{10} cycles. d) Comparison of the E_{bias} evolution for Sr:HfO$_2$ based FeRAM and ZrO$_2$ based AFE-RAM. Yellow denotes maximum Preisach density whereas blue color marks the non-switching regions.

Beside the distribution of the coercive field that influences the variability of the device, for low power operation, a low magnitude of the coercive field is needed. Compared to ferroelectric Sr:HfO$_2$, the coercive field of the AFE-RAM is reduces to less than half.

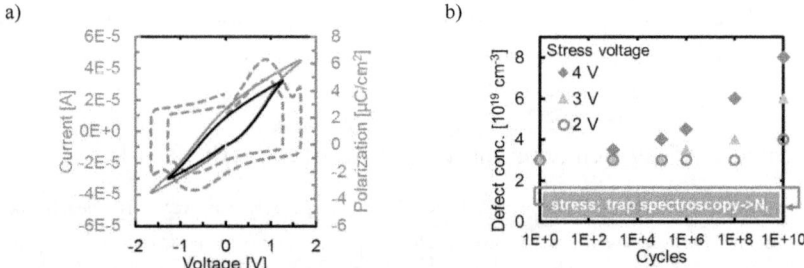

Figure 7.12 Sub-cycling operation of AFE-RAM. a) I-V and corresponding P-V characteristics together with b) defect concentration evolution with field cycling recorded for different stress conditions. Figure taken from study published by Pešić et al. [77].

7 ANTI-FERROELECTRIC NON-VOLATILE MEMORY

Further, in order to use the material as a multi-level cell memory, one of the most important properties of the ferroelectric memories is the sub-cycling operation. Accordingly, sub-cycling tests were performed with operating conditions as low as 1.3 V (see Figure 7.12a). Such remarkable low voltage performance is possible precisely because of the very low magnitude and narrow distribution of the coercive/critical field of 0.7 MV/cm within the polycrystalline stack of the AFE-RAM. The use of low operation fields/cycling stress conditions results in a very stable endurance with minimal degradation of the stack. Different stressing PRG/ERS voltages were used for recording of the defect concentration evolution. Again, trap defect spectroscopy (see details in section 4.1) was utilized for extraction of defect concentration from stress dependent leakage current characteristics (7.12b). As expected, the increase of the stressing amplitude resulted in earlier onset and stronger generation of the electrically active defects.

7.5.3 Exploring the Limits of ZrO_2-based AFE-Memories

Due to the similarity of the AFE-RAM stack to standard DRAM, a similar trick for reliability improvement and further leakage current reduction can be used. As discussed before in Chapter 5, an Al_2O_3 interlayer introduced between the two sandwiching ZrO_2 layers blocks the propagation of grain boundaries from the top to the bottom electrode, significantly reducing the leakage current and increasing the reliability. Guided by this approach, $TiN/ZrO_2/Al_2O_3/ZrO_2$ stack were fabricated with TiN and RuO_x TE (see Chapter 3.1 and Figure 3.2b for details). As expected, the leakage currents of the devices were reduced, while the operation voltage was only slightly increased (Figure 7.13a). This is due to the fact that the amorphous top ZrO_2 layer and Al_2O_3 interlayer represent a dead-layer and series capacitance with lower k-value (just 9 – 10 instead of 30 for ZrO_2) causing a voltage drop as well as a certain depolarization field.

The use of only half of the AFE-double-loop, together with the low voltage operation and series resistance due to the blocking interlayer results in a rather small P_r value which reduces the read margin (see Chapter 2.5). Thinking of highly scaled large memory arrays and significant bit line capacitances, an appropriate discrimination between the logical states 1 and 0 becomes more and more challenging if a certain difference in the amount of charge flowing to the bit line during the read operation of the respective cannot be maintained. This can be done by increasing the capacitor area if all other parameters are kept fixed. This approach is common for state-of-the-art DRAM and is based on the realization of 3D capacitors to keep the footprint of the memory cell as small as possible. Therefore, the possibility of implementing AFE-RAM capacitors in a 3D integrated fashion was checked. Guided by the observed behavior for the planar ZrO_2 films, standard 3D DRAM capacitor structures comprising ZAZ stack provided by Qimonda [93] was examined (see details in Section 3.1). Measurements were performed on fully integrated test structures with 3D capacitors having an aspect ratio of 30:1 (see Figure7.13c inset). These 3D structures were realized in

7 ANTI-FERROELECTRIC NON-VOLATILE MEMORY

a 46 nm 6F^2 DRAM technology. Details about the DRAM structures are reported elsewhere [93]. TEM cross-sections of these capacitors fully integrated buried word-line technology are shown in Figure 7.13c. A comparison of the planar and 3D structures is given in Figure 7.13b. It can be seen that 3D integration results in about 30 times higher polarization values per projected area with respect to the planar structures. As recently reported by Pešić et al., the obtained results together with availability of a RuO$_x$ ALD process [191] paves the way for the feasible integration of the AFE-RAM into 3D structures providing a sufficiently large MW for a 1T-1C memory architecture. This makes this novel concept a serious competitor for today's DRAM generations, offering similar scaling capabilities coupled with non-volatile storage, i.e. a promising candidate for AFE-RAM.

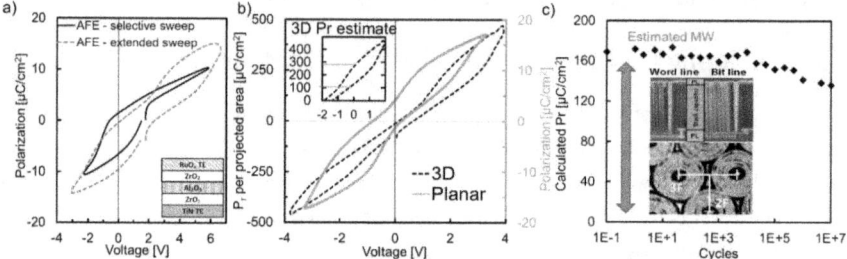

Figure 7.13 a) P-V characteristics of ZAZ based AFE-RAM. b) P-V curve for a 2D (right axis, solid) and a 3D (left axis, dashed) ZrO$_2$ based capacitor. (Inset) estimated polarization gain obtained by 3D integration of a TiN/ZAZ/TiN structure exhibiting AFE behavior. c) Endurance of a 3D ZAZ capacitor recorded with ±4 V at 300 kHz. Inset SEM cross-sections and TEM top-view micrograph of integrated 3D capacitors. Figure taken from study published by Pešić et al. [77].

Having in mind this excellent performance of AFE-RAM in possible 1T-1C designs, the realization of the proposed concept in 1T architectures represents the next logical step. In ferroelectric memories, the transition from capacitor to transistor based storage offers the possibility to spare the rather large capacitor and reduce the foot print of the cell tremendously, to allow the implementation of memories with even higher storage density. Besides a higher storage density, 1T architecture enables higher scalability and most importantly non-destructive read-out. To this aim, the previously discussed stack comprising /ZrO$_2$/Al$_2$O$_3$/ZrO$_2$/RuO$_x$ is deposited on the medium doped p-type silicon substrates with SiO$_2$ interface buffer layers (Figures 7.14a-b). Performing the previously discussed patterning of the gate electrodes, AFE-MIS capacitors were formed. Applying the triangular voltage excitation, an I-V characteristic characterized with two switching peaks was obtained (Figure 7.14c). Similar to the MFIS structure discussed in Section 6.4, voltage drops over the semiconductor as well as over the interface buffer layer shifted the peaks towards higher electric fields. Analogous to the AFE-RAM device, a wake-up-less I-V and corresponding P-V characteristics were obtained. Even though the lower E_c implies a lower MW, this initial MIS study shows the potential to overcome the wake-up issues of hafnia based FeFETs shown in Section 6.4 via the AFE-FET concept. Further it should be noted that as in the FeFET case, the interface buffer layer remains a weak link and limiting constraint due to the high fields over this low-k material. Nonetheless, the wake-up free endurance and stability of AFE-based FETs would enable implementation of the AFE-FET in the embedded NVM solutions.

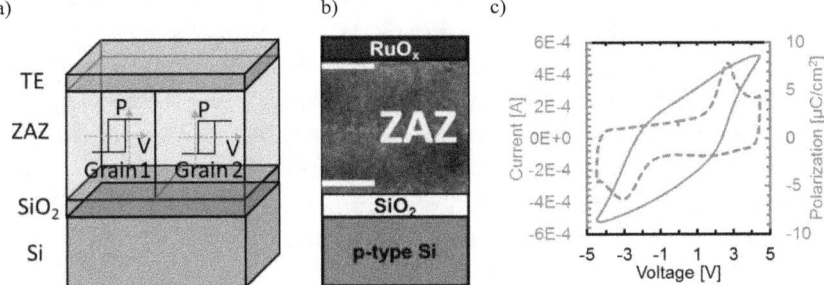

Figure 7.14 MOS realization of the AFE-FET concept. a) Gate stack comprising $Si/SiO_2/ZrO_2/Al_2O_3/ZrO_2/RuO_x$ and b) corresponding TEM micrograph of the ZAZ layer c) Pristine I-V and P-V characteristics of the AFE-MOS capacitor. Figure taken from study published by Pešić et al. [77].

7.5.4 Temperature Stability of the AFE-based Non-volatile Memories

As discussed before, since of the beginning of utilization of ferroelectric films as in non-volatile memory devices, imprint is considered to be the most critical property that, besides the issues of scalability, has hindered the final application and production of various ferroelectric-like materials [14]. Imprint is generally caused by diffusion of oxygen vacancies and electron charging of defects.

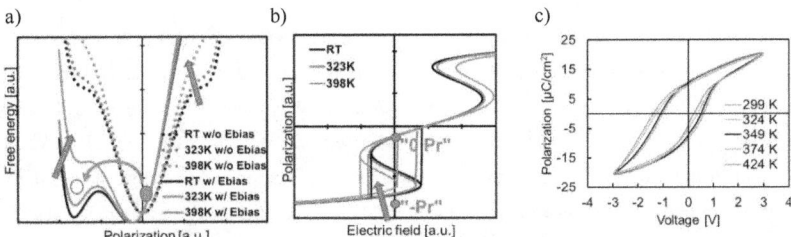

Figure 7.15 a) Simulation of the temperature dependence of the Landau potential and b) consequent shift of the hysteresis characteristics with increase of the temperature. c) Temperature dependent P-V characteristics of the MIM AFE-RAM device.

In order to explore the temperature stability of the suggested stack, the LK model described in Section 2.3 was utilized to simulate the temperature influence on the free energy potential landscape. In Figure 7.15a it can be seen that with the increase of the operating temperature, there is a strong decrease of the side minima which reduces the effective field needed for the backswitching to the unpolarized state of the device. Consequently, the P-V characteristic (Figure 7.15b) is being shifted towards lower field (imprinted) with increasing temperature, which results in a reduction, and at certain point, a complete closure of the MW. In order to verify if this issue is more critical than in the FE based memories studied in Chapter 6, temperature dependent P-V experiments (Figure 7.15c) were performed. The experiment resulted in strong imprint with increasing temperature and thus, verified the simulation outcome in Figure 7.15a-b. In between each dynamic hysteresis measurement, a DUT temperature was sequentially increased. Afterwards, E_c of the stack was extracted from the P-V characteristics obtained by DHM and plotted against temperature. The experiment performed on the ZrO_2 based MIM structure confirmed the observation suggested by the model. Indeed, with increase of the temperature the P-V characteristics are shifted towards lower fields whereas at 125 °C and above the MW was strongly decreased (see Figure 7.16a).

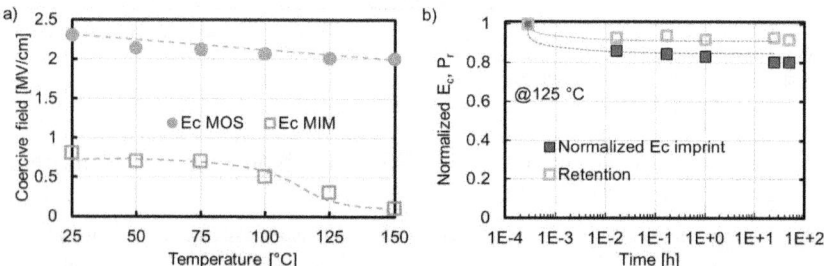

Figure 7.16 a) Temperature stability of an AFE-MIS capacitor in comparison with an AFE MIM structure tested at 25 – 150 °C. MW of the MIS can be approximated by MW~ $2E_c d$ where E_c is the coercive field and d is the thickness of the layer. b) Retention and imprint of a $Si/SiO_2/ZAZ/TiN$ AFE-MIS structure. Figure taken from study published by Pešić et al. [77].

Moving from MIM stacks (as of interest for 1T-1C AFE memories) to MIS stacks (as used in 1T AFE memory cells), the Si substrate instead of TiN serves as bottom electrode. This should increase the asymmetry of the stack and thus, increase the barrier governing the stability of the two states. To address this question, a temperature dependent study was performed on the ZrO_2 based AFE-MIS structures. Initially, similar to those just discussed, imprint experiments were performed (In between each dynamic hysteresis measurements, the sample was left for bake-out for different times, after which E_c was recorded). Even though imprint was observable, AFE-MIS capacitor exhibited a rather small shift with respect to the

7 ANTI-FERROELECTRIC NON-VOLATILE MEMORY

ZrO_2 based MIM device. In contrast to the MIM structure that loses its state at 125 °C, the AFE-MIS devices exhibited stable retention at 125 °C for more than a 100 h.

7.6 Summary

Within this chapter, the possibility of using modified anti-ferroelectric materials characterized by higher breakdown strength and much higher endurances, as new non-volatile materials is considered. Guided by the studies that reported that tetragonal phase as responsible for the anti-ferroelectric behavior, a stabilization of the tetragonal phase was induced in both PVD and ALD deposited film. This tetragonal phase revealed never before reported AFE properties of standard DRAM capacitor stacks. Having stabilized the desired phase, the conceptual problem of the anti-ferroelectric was addressed. Namely, anti-ferroelectric materials cannot be directly utilized as memory medium due to the already mentioned loss of the memory state upon removal of the external field. To address this issue, it was shown that introduction of an internal bias field is needed for centering of one of the AFE loops and providing memory capabilities known for ferroelectric materials. The needed internal bias field could be induced by asymmetric electrodes with different workfunction values. To analyze the capabilities of these materials and to offer a proof of concept, a Landau-Ginzburg-Devonshire formalism was introduced. Detailed simulations confirmed the proposed concept of an AFE-RAM that uses a polarized and depolarized state for distinguishing between logical "1" and logical "0". The internal bias was created by introduction of a high workfunction RuO_x top electrode, thus enabling the AFE-like nonvolatile memory. Utilizing this feature, the exceptional endurance of anti-ferroelectrics was demonstrated and made a state-of-the-art DRAM stack non-volatile. In addition to the 1T-1C concept, a proof of concept for a 1T architecture based on AFEMIS devices was developed. Detailed electrical characterization showed high endurance and retention values of this novel non-volatile concept.

8 Summary and Outlook

With the internet of things market expansion and tremendous explosion of the amount of data generated in last years, computer systems are shifting from computing oriented to data oriented systems. Having in mind that all data centers in the world consume almost 7% of world's power, the necessity for the low power fast solution grows. In order to improve the performance of the whole system, the mismatch between logic and memory has to be decreased. To address this issue, a rather large variety of emerging memories are currently investigated. Among all, PZT-based ferroelectric memory offers nanosecond operation range, almost unlimited endurance, non-volatility and thus promotes itself as a logical solution. Despite the effort invested by various groups, the main obstacle of integration of this from CMOS point of view exotic material into the state-of-the art devices persists.

The discovery of the ferroelectricity in HfO_2 enabled overcoming the integration and compatibility issues and bridged the gap between the state-of-the-art semiconductor processing and ferroelectric memories. Best examples are ultimately scaled FeFET fabricated in 28 nm HKMG technology, as well as the 3D integration of the ferroelectric hafnia into the capacitor structures with high aspect ratios. Even though both FeRAM and FeFET offer remarkable performance and nominate themselves as potential candidates for a storage class memory, both memory architectures suffer from limited endurance. To address this fundamental requirement and understand the cause of limited endurance this thesis was carried through three consecutive stages built atop of each other.

Studying influence of field cycling on the thin films with spontaneous polarization is extremely complex. A physical interplay governs the purely dielectric mechanisms and ferroelectric properties. With that in mind, the complexity of this strongly entangled material system was decreased in Chapter 5. This was necessary to get an idea on how defects, charged defects and pure dielectric engineering influence the leakage current and further ferroelectric properties within the film. These mechanisms are connected to each other and take place both at interfacial and in bulk regions. As a test structure ZrO_2-based spontaneous polarization-free capacitors were chosen. ZrO_2 was chosen due to it is similarity with HfO_2, as well as the fact that approaches used for improvement of the DRAM stack can be applied to more complex ferroelectric stacks. Initially, to address interface properties and their influence on the charge transport and reliability, different top electrode materials X (Pt, TiN) were compared in a $TiN/ZrO_2/Al_2O_3/ZrO_2/X$ MIM capacitor structure. Pt led to improved leakage current behavior compared to TiN electrodes. In addition, this noble metal electrode material showed an improvement in capacitor reliability with respect to the TiN top electrode. Detailed parameter extraction led to development of the model describing the charge transport

8 SUMMARY AND OUTLOOK

within the ZAZ-based MIM stacks. Afterwards, the carefully evaluated charge transport model was extended and applied to the ferroelectric, Sr doped HfO_2 system. The model was implemented in TCAD in order to address the interplay of the trapping and ferroelectric switching process. In parallel to the TCAD modeling the charge transport model was transferred to MDLab dielectric simulator to prepare the model that will be used to simulate the oxygen vacancy drift/diffusion processes due to field cycling and its influence on the switching characteristics of the ferroelectric.

After the detailed analysis of the charge transport and determining parameters responsible for the reliability of the stack, an extension of the previously developed charge transport model of ferroelectric capacitor is performed in Chapter 6 to address central reliability issues of ferroelectric-based memories. Field cycling stability represents one of the major requirements of a memory. Within this chapter, the mechanisms responsible for the field cycling behavior of HfO_2 based ferroelectric capacitors were studied. Two main stages of the devices' lifetime, i.e., wake-up and fatigue, were investigated. A combination of comprehensive experimental methods was used to identify that no defects are generated but rather already existing defects are redistributed within the device in the initial stage of lifetime. The faster diminishing of the internal bias field at elevated temperatures solidified the assumption that the removal of internal bias field, i.e. the creation of the uniform field distribution as well as the phase transition, during the wake-up was diffusion and drift driven. In addition, a TEM study suggested a phase transition of the interface regions during cycling. These results motivated comprehensive modeling of the vacancies, ion movement, as well as their recombination within the stack. By combining the modeling of the diffusion mechanisms and phase transition, it was shown that both are responsible for increase of remnant polarization during the first stage of the device lifetime (wake-up). Furthermore, the root cause of the limited endurance of the doped HfO_2-based ferroelectric MIMs was identified to be the dielectric degradation which reduces the ferroelectric switching. The increase of the defect density with field cycling indicates that the main mechanism responsible for the degradation of the ferroelectric behavior besides domain pinning is defect generation. Electron trapping at these defects affects the field distribution within the stack reducing the field in the bulk of the ferroelectric layer. A further generation of vacancies creates leakage paths, finally resulting in a breakdown of the stack before the memory window closes completely.

From the presented, it can be concluded that internal bias fields and field driven phase transition are responsible for the wake-up (increase of the remnant polarization) of ferroelectric memories in the initial parts of the lifetime. Besides the high fields needed for reaching of the coercive field and altering of the polarization are in the range of the breakdown of the dielectric, thus indicating that examined device has a rather conceptual problem. One solution would be to find a novel dopant that would reduce the coercive field and operation voltage and thus enhance the lifetime of the device. Another solution would be

8 SUMMARY AND OUTLOOK

utilization of the material with higher endurance strength. Such materials were proven to be anti-ferroelectric binary oxides. In contrast to the ferroelectric, anti-ferroelectric materials lose their polarization with removal of the external field. To address this conceptual problem an intentionally asymmetric stack was presented. In Chapter 7 it is shown that introduction of an internal bias field is needed for centering of one of the AFE loops and providing memory capabilities known for ferroelectric materials. The needed internal bias field could be induced by asymmetric electrodes, with different workfunction values. To analyze the feasibility and offer a proof of concept, Landau-Ginzburg-Devonshire formalism was introduced. Detailed simulations confirmed the proposed concept of an AFE-RAM that uses a polarized and depolarized state for distinguishing between logical "1" and logical "0". The internal bias was created by introduction of a high workfunction RuO_x top electrode, thus enabling the AFE-like nonvolatile memory. With such approach a standard ZrO_2 based DRAM stack was made non-volatile and world's first AFE-RAM was demonstrated. Besides the 1T-1C concept, a proof of concept for a 1T architecture based on AFEMIS devices was offered. Detailed electrical characterization showed high endurance and retention values. By providing the speed of FE memories while enabling stable and low power operation, discovery of the AFE-RAM paves the way for a new paradigm for universal memories.

Having in mind that the drift and diffusion are governed by the electric field and temperature, respectively, the integration questions should be carefully crosschecked for possible material diffusion which may take place due to the thermal treatment as well as due to generated field within the asymmetric stack. The presented findings and superior properties of the anti-ferroelectric shall motivate in-depth reliability studies of their high-endurance strength as well as the structural analysis behind the field-induced phase transitions which might be responsible for the anti-ferroelectric-like properties of the stack.

Bibliography

[1] C. S. Hwang et al., "Prospective of Semiconductor Memory Devices: from Memory System to Materials," *Adv. Electron. Mater.*, vol. 1, no. 6, pp. 1–30, 2015.

[2] J. Gantz et al., "THE DIGITAL UNIVERSE IN 2020: Big Data, Bigger Digital Shadows, and Biggest Growth in the Far East," *Idc*, vol. 2007, no. December 2012, pp. 1–16, 2012.

[3] F. M. Wanlass et al., "Low stand-by power complementary field-effect circuitary," 1957.

[4] P. Ranganathan et al., "From microprocessors to nanostores: Rethinking data-centric systems," *Computer (Long. Beach. Calif).*, vol. 44, no. 1, pp. 39–48, 2011.

[5] M. Qazi et al., "Challenges and directions for low-voltage SRAM," *IEEE Des. Test Comput.*, vol. 28, no. 1, pp. 32–43, 2011.

[6] A. Chen et al., "A review of emerging non-volatile memory (NVM) technologies and applications," *Solid. State. Electron.*, vol. 125, pp. 25–38, 2016.

[7] N. Nagel et al., "Emerging Non-Volatile Memory Technologies," pp. 37–44, 1995.

[8] Y. Xie et al., *Emerging memory technologies: Design, architecture, and applications*, vol. 9781441995. 2014.

[9] D. Apalkov et al., "Spin-transfer torque magnetic random access memory (STT-MRAM)," *ACM J. Emerg. Technol. Comput. Syst.*, vol. 9, no. 2, p. 13:1-13:35, 2013.

[10] D. S. Jeong et al., "Emerging memories: resistive switching mechanisms and current status," *Reports Prog. Phys.*, vol. 75, p. 76502, 2012.

[11] R. Waser et al., "Nanoionics-based resistive switching memories.," *Nat. Mater.*, vol. 6, no. 11, pp. 833–40, 2007.

[12] R. F. Freitas et al., "Storage-class memory: The next storage system technology," *IBM J. Res. Dev.*, vol. 52, no. 4.5, pp. 439–447, 2008.

[13] U. Bottger et al., "Ferroelectric Random Access Memories," *Nanoelectron. Inf. Technol.*, vol. 12, no. 10, pp. 565–590, 2003.

[14] J. F. Scott et al., "Ferroelectric Memories," *Science (80-.).*, vol. 246, no. 4936, pp. 213–1405, 1989.

[15] J. Muller et al., "Ferroelectric hafnium oxide: A CMOS-compatible and highly scalable approach to future ferroelectric memories," in *Technical Digest - International Electron Devices Meeting, IEDM*, 2013.

[16] "Emerging non-volatile memories," *IEEE Solid-State Circuits Magazine*, vol. 8, no. 2, p. 44, 2016.

[17] E. Yurchuk et al., "Electrical Characterisation of Ferroelectric Field Effect Transistors based on Ferroelectric HfO_2 Thin Films," 2015.

[18] U. Schroeder et al., "Hafnium Oxide Based CMOS Compatible Ferroelectric Materials," *ECS J. Solid State Sci. Technol.*, vol. 2, no. 4, pp. N69–N72, 2013.

[19] S. Mueller et al., "Development of HfO_2-Based Ferroelectric Memories for Future CMOS Technology Nodes by," 2014.

[20] H. Kohlstedt et al., "Current status and challenges of ferroelectric memory devices," *Microelectron. Eng.*, vol. 80, no. SUPPL., pp. 296–304, 2005.

[21] M. Pešić et al., "Root cause of degradation in novel HfO_2-based Ferroelectric Memories," *IEEE*, 2016.

[22] M. Pešić et al., "Physical Mechanisms behind the Field-Cycling Behavior of HfO_2 -Based Ferroelectric Capacitors," *Adv. Funct. Mater.*, vol. 26, pp. 4601–4612, 2016.

[23] M. T. Bohr et al., "The high-k solution," *IEEE Spectr.*, vol. 44, no. 10, pp. 29–35, 2007.

[24] K. Tse et al., "Oxygen vacancies in high-k oxides," *Microelectron. Eng.*, vol. 84, no. 9–10, pp. 2028–2031, 2007.

[25] J. Robertson et al., "Defect Energy Levels in High-K Gate Oxides," *Defects High-k Gate Dielectr. Stacks*, Springer, pp. 175–187, 2006.

[26] J. Robertson et al., "High dielectric constant oxides," *Eur. Phys. J. Appl. Phys.*, vol. 28, pp. 265–291, 2004.

[27] R. Waser et al., "Nanoelectronics and Information Technology, 3rd Edition," John Wiley & Sons, p. 1040, 2012.

[28] V. V. Afanas'ev et al., "Band alignment at the interfaces of Al_2O_3 and ZrO_2 -based insulators with metals and Si," *J. Non. Cryst. Solids*, vol. 303, pp. 69–77, 2002.

[29] D. K. Schroder et al., "Semiconductor material and device characterization". *Third Edition*, John Wiley & Sons vol. 44, no. 4. 2006.

[30] J. Robertson et al., "Band offsets of wide-band-gap oxides and implications for future electronic devices," *J. Vac. Sci. Technol. B Microelectron. Nanom. Struct.*, vol. 18, no. 3, p. 1785, 2000.

[31] C. A. Aita et al., "Reactive sputter deposition of metal oxide nanolaminates," *J. Phys. Condens. Matter*, vol. 20, no. 26, pp. 8–10, 2008.

[32] J. E. Jaffe et al., "Low-temperature polymorphs of ZrO_2 and HfO_2: A density-functional theory study," *Phys. Rev. B - Condens. Matter Mater. Phys.*, vol. 72, no. 14, pp. 1–9, 2005.

[33] X. Zhao et al., "First-principles study of structural, vibrational, and lattice dielectric properties of hafnium oxide," *Phys. Rev. B*, vol. 65, no. 23, pp. 75105-1–10, 2002.

[34] C. Adelmann et al., "Atomic Layer Deposition of Gd-Doped HfO_2 Thin Films," *J. Electrochem. Soc.*, vol. 157, no. 4, p. G105, 2010.

[35] V. A. Gritsenko et al., "Electronic properties of hafnium oxide: A contribution from defects and traps," *Phys. Rep.*, vol. 613, pp. 1–20, 2016.

[36] S. Clima et al., "First-principles simulation of oxygen diffusion in HfO_x: Role in the resistive switching mechanism," *Appl. Phys. Lett.*, vol. 100, no. 13, pp. 3–7, 2012.

[37] R. Degraeve et al., "Dynamic 'hour glass' model for SET and RESET in HfO_2 RRAM," *Dig. Tech. Pap. - Symp. VLSI Technol.*, no. June 2016, pp. 75–76, 2012.

[38] L. Larcher et al., "Microscopic understanding and modeling of HfO_2 RRAM device physics," *Tech. Dig. - Int. Electron Devices Meet. IEDM*, pp. 474–477, 2012.

[39] T. S. Böscke et al., "Ferroelectricity in hafnium oxide thin films Ferroelectricity in hafnium oxide thin films," *Appl. Phys. Lett.*, vol. 102903, pp. 0–3, 2011.

[40] M. Hoffmann et al., "Direct Observation of Negative Capacitance in Polycrystalline Ferroelectric HfO$_2$," *Adv. Funct. Mater.*, 2016.

[41] M. H. Lee et al., "Prospects for Ferroelectric HfZrOx FETs with Experimentally CET = 0.98nm , SS for = 42mV / dec , SS rev = 28mV / dec , Switch-OFF < 0.2V , and Hysteresis-Free Strategies," *Int. Electron Devices Meet.*, pp. 616–619, 2015.

[42] M. Pešić et al., "Low leakage ZrO$_2$ based capacitors for sub 20 nm dynamic random access memory technology nodes," *J. Appl. Phys.*, vol. 119, no. 6, Feb. 2016.

[43] M. Pešić et al., "Conduction barrier offset engineering for DRAM capacitor scaling," *Solid. State. Electron.*, vol. 115, pp. 133–139, Jan. 2016.

[44] K. Cho et al., "Schottky barrier height engineering for next generation DRAM capacitors," in *Joint International EUROSOI Workshop and International Conference on Ultimate Integration on Silicon (EUROSOI-ULIS)*, pp. 129–132, 2015.

[45] J. Muller et al., "Ferroelectric hafnium oxide : A CMOS - compatible and highly scalable approach to future ferroelectric memories," in *IEEE International Electron Devices Meeting (IEDM)*, p. 10.8.1-10.8.4, 2013.

[46] J. Müller et al., "Ferroelectricity in HfO$_2$ enables nonvolatile data storage in 28 nm HKMG," in *IEEE 2012 Symposium onVLSI Technology (VLSIT)*, 2012, no. 0.

[47] T. V Perevalov et al., "Origin of defects responsible for charge transport in HfO$_2$," no. July, pp. 3–4, 2016.

[48] E. Gusev et al. "Defects in High-k Gate Dielectric Stacks" Springer NATO Science Series 2006.

[49] G. C. Jegert et al., "Modeling of Leakage Currents in High-k Dielectrics," 2011.

[50] F. Chiu et al., "A Review on Conduction Mechanisms in Dielectric Films," Advances in Materials Science and Engineering, 2014.

[51] L. Larcher et al., "Statistical simulation of leakage currents in MOS and flash memory devices with a new multiphonon trap-assisted tunneling model," *IEEE Trans. Electron Devices*, vol. 50, no. 5, pp. 1246–1253, 2003.

[52] R. M. Hill et al., "Poole-Frenkel conduction in amorphous solids," *Philos. Mag.*, vol. 23, no. 181, pp. 59–86, 1971.

[53] K. A. Nasyrov et al., "Charge transport in dielectrics via tunneling between traps," *J. Appl. Phys.*, vol. 109, no. 9, pp. 1–6, 2011.

[54] L-2016.03, "Sentaurus ^{TM} Device User," no. September, 2014.

[55] L. Vandelli et al., "A physical model of the temperature dependence of the current through SiO$_2$/HfO$_2$ stacks," *IEEE Trans. Electron Devices*, vol. 58, no. 9, pp. 2878–2887, 2011.

[56] L. Vandelli et al., "Modeling temperature dependency (6 – 400K) of the leakage current through the SiO$_2$/high-k stacks," *2010 Proc. Eur. Solid State Device Res. Conf.*, pp. 388–391, 2010.

[57] R. Tsu et al., "Tunneling in a finite superlattice," *Appl. Phys. Lett.*, vol. 22, no. 11, pp. 562–564, 1973.

[58] C. Kittel et al., "Theory of Antiferroelectric Crystals," Physical Review, vol. 82, no. 729, 1951.

[59] A. K. Tagantsev et al., "The origin of antiferroelectricity in PbZrO$_3$.," *Nat. Commun.*, vol. 4, p.

2229, 2013.

[60] S.-E. Park et al., "Electric field induced phase transition of antiferroelectric lead lanthanum zirconate titanate stannate ceramics," *J. Appl. Phys.*, vol. 82, no. 4, p. 1798, 1997.

[61] M. Pešić et al., "Nonvolatile Random Access Memory and Energy Storage Based on Antiferroelectric Like Hysteresis in ZrO_2," *Adv. Funct. Mater.*, 26: 7486–7494. doi:10.1002/adfm.201603182.

[62] U. Böttger, "1 Dielectric Properties of Polar Oxides," Polar Oxides: Properties, Characterization, and Imaging, 11-38. 2005.

[63] P. Chandra and P. B. Littlewood, "A Landau primer for ferroelectrics," Springer Berlin Heidelberg, pp. 69–116, 2007.

[64] K. Dragosits et al., "Transient Simulation of Ferroelectric Hysteresis," Proc. 3rd Int. Conf. on Modeling and Simulation of Microsystems, pp. 433–436, 2000.

[65] S. Mueller et al., "Performance Investigation and Optimization of Si:HfO_2 FeFETs on a 28 nm Bulk Technology," IEEE International Symposium on the Applications of Ferroelectric and Workshop on the Piezoresponse Force Microscopy (ISAF/PFM), (pp. 248-251) 2013.

[66] G. Fox et al., "Thin Films Texture and Scaling Effects in Ferroelectric Random Access Memory," in *Gordon Research Conference (Ceramics, Solid State Studies)*.

[67] K. Tomida et al., "Dielectric constant enhancement due to Si incorporation into HfO_2," *Appl. Phys. Lett.*, vol. 89, no. 14, pp. 1–4, 2006.

[68] A. Toriumi et al., "Doped HfO_2 For Higher-k Dielectric," *ECS Trans.*, vol. 1, no. 5, pp. 185–197, 2006.

[69] R. Materlik et al., "The origin of ferroelectricity in $Hf_{1-x}Zr_xO_2$: A computational investigation and a surface energy model," *J. Appl. Phys.*, vol. 117, no. 13, p. 134109, 2015.

[70] T. Schenk et al., "Formation of Ferroelectricity in Hafnium Oxide Based Thin Films," PhD Thesis *TU Dresden*, 2017.

[71] X. Sang et al., "On the structural origins of ferroelectricity in HfO2 thin films On the structural origins of ferroelectricity in HfO_2 thin films," *Appl. Phys. Lett.*, vol. 106, no. 105, pp. 162905–61614, 2015.

[72] R. D. Shannon et al., "Revised effective ionic Radii and systematic studies of interatomic distances in Halides and Chalcogenides," *Acta Cryst.*, vol. A32, pp. 751–767, 1976.

[73] M. H. Park et al., "Ferroelectricity and Antiferroelectricity of Doped Thin HfO_2-Based Films," *Adv. Mater.*, vol. 27, no. 11, pp. 1811–1831, 2015.

[74] X. J. Lou et al., "Why do antiferroelectrics show higher fatigue resistance than ferroelectrics under bipolar electrical cycling?," *Appl. Phys. Lett.*, vol. 94, no. 7, pp. 2007–2010, 2009.

[75] L. Zhou et al., "Electric fatigue in antiferroelectric $Pb_{0.97}La_{0.02}(Zr_{0.55}Sn_{0.33}Ti)O_3$ ceramics induced by bipolar cycling," *J. Eur. Ceram. Soc.*, vol. 26, no. 6, pp. 883–889, 2006.

[76] U. Schroeder et al., "Impact of different dopant on the witching propertie of ferroelectric hafniumoxide," *Jpn. J. Appl. Phys.*, vol. 63, no. 8S1, pp. 8–10, 2014.

[77] M. Pešić et al., "How to make DRAM non-volatile? Anti- ferroelectrics: A new paradigm for

universal memories," *016 IEEE International Electron Devices Meeting (IEDM)*, San Francisco, CA, 2016, pp. 11.6.1-11.6.4..

[78] D. A. Buck *et al.*, "'Ferroelectrics for digital information storage and switching,' master thesis, MIT Digital Computer Laboratory (1952)," 1952.

[79] J. R. Anderson *et al.*, "Ferroelectric materials as storage elements for digital computers and switching systems," *Trans. Am. Inst. Elect. Eng. 71, Part I Commun.Electron*, pp. 395–401, 1953.

[80] Ramtron. F-ram technology brief. Technical report, Ramtron, 2007. URL: "http://www.digikey.com/- Web%20Export/Supplier%20Content/ramtron-1140/pdf/ramtron-tech-ferroelectric.pdf?redirected=1.," p. 2007, 2007.

[81] Texas Instruments, "Texas Instruments Delivers First Chip Made on Advanced 90nm Process," *http://newscenter.ti.com/news-releases?item=126318*, 2003.

[82] S. R. Summerfelt *et al.*, "High-density 8Mb 1T-1C ferroelectric random access memory embedded within a low-power 130 nm logic process," *IEEE Int. Symp. Appl. Ferroelectr.*, pp. 9–10, 2007.

[83] J. Müller *et al.*, "Ferroelectric Hafnium Oxide A Game Changer to FRAM?," in *Non-Volatile Memory Technology Symposium (NVMTS), 2014 14th Annual*, pp. 1–7, 2014.

[84] T. Mikolajick *et al.*, "Doped Hafnium Oxide – An Enabler for Ferroelectric Field Effect Transistors," *Adv. Sci. Technol.*, vol. 95, pp. 136–145, 2014.

[85] International Technology Roadmap for Semiconductors (ITRS), "Emerging Research Materials (ERM)," pp. 1–98, 2013.

[86] U. Schroeder *et al.*, "Materials for DRAM Memory Cell Applications," in "Thin Films on Silicon: Electronic and Photonic Applications", 369, 2016.

[87] P. T. Mikolajick *et al.*, "Memory Technology II Chapter 2 – Ferroelectric Memories Part II : Design and Operation , Reliability and Status," 2015.

[88] C.-P. Yeh, *et al.*, "Fabrication and investigation of three-dimensional ferroelectric capacitors for the application of FeRAM," *AIP Adv.*, vol. 6, no. 3, p. 35128, 2016.

[89] ITRS, "International Technology Roadmap for Semiconductors-Process Integration Devices and Structures," *ITRS Publ.*, 2011.

[90] P. Polakowski *et al.*, "Ferroelectric deep trench capacitors based on Al:HfO$_2$ for 3D nonvolatile memory applications," *2014 IEEE 6th Int. Mem. Work. IMW 2014*, pp. 1–4, 2014.

[91] J. Müller *et al.*, "Ferroelectricity in HfO$_2$ enables nonvolatile data storage in 28 nm HKMG," pp. 25–26, 2012.

[92] E. Yurchuk *et al.*, "Impact of Scaling on the Performance of HfO$_2$ - Based Ferroelectric Field Effect Transistors," vol. 61, no. 11, pp. 3699–3706, 2014.

[93] T. Schloesser *et al.*, "6F^2 buried wordline DRAM cell for 40 nm and beyond," in *IEEE International Electron Devices Meeting* (IEDM), *2008*, pp. 2–4, 2008.

[94] T. Schenk *et al.*, "Electric field cycling behavior of ferroelectric hafnium oxide," *ACS Appl. Mater. Interfaces*, vol. 6, no. 22, pp. 19744–19751, Nov. 2014.

[95] J. Yuan *et al.*, "Performance elements for 28 nm gate length bulk devices with gate first high-k metal gate," in *10th IEEE* International Conference on Solid-State and Integrated Circuit Technology

(ICSICT), pp. 66-69, 2010.

[96] G. Gai Gianni *et al.*, "Study of leakage mechanism and trap density in porous low-k materials," in *IEEE International Reliability Physics Symposium Proceedings (IRPS)*, pp. 549-555, 2010.

[97] M. Vilmay *et al.*, "Characterization of low-k SiOCH dielectric for 45 nm technology and link between the dominant leakage path and the breakdown localization," *Microelectron. Eng.*, vol. 85, no. 10, pp. 2075–2078, 2008.

[98] M. Kerber *et al.*, "Trap related dielectric absorption of HfSiO films in metal-insulator-semiconductor structures," *J. Vac. Sci. Technol. B Microelectron. Nanom. Struct.*, vol. 27, no. 1, p. 321, 2009.

[99] H. Reisinger *et al.*, "A comparative study of dielectric relaxation losses in alternative dielectrics," *IEEE Int. Electron Devices Meet.*, p. 12.2.1-12.2.4, 2001.

[100] M. Schumacher *et al.*, "Dielectric relaxation phenomena in superparaelectric and ferroelectric ceramic thin films and the relevance with respect to high density DRAM and FRAM applications," *J. Phys. Iv*, vol. 8, pp. 117–120, 1998.

[101] M. Duschl *et al.*, "Reliability aspects of Hf-based capacitors : Breakdown and trapping effects," *Microelectron. Reliab.*, vol. 47, no. 4–5, pp. 497–500, 2007.

[102] C. B. Sawyer *et al.*, "Rochelle salt as a dielectric," *Phys. Rev.*, vol. 35, no. 3, pp. 269–273, 1930.

[103] C. R. Pike *et al.*, "Characterizing interactions in fine magnetic particle systems using first order reversal curves," *J. Appl. Phys.*, vol. 85, no. 9, pp. 6660–6667, 1999.

[104] C. R. Pike *et al.*, "First order reversal curve diagrams and thermal relaxation effects in magnetic particles," *Geophys. J. Int.*, vol. 145, no. 3, pp. 721–730, 2001.

[105] L. Cima *et al.*, "Characterization and model of ferroelectrics based on experimental Preisach density," *Rev. Sci. Instrum.*, vol. 73, no. 10, p. 3546, 2002.

[106] A. Stancu *et al.*, "First-order reversal curves diagrams for the characterization of ferroelectric switching," *Appl. Phys. Lett.*, vol. 83, no. 18, pp. 3767–3769, 2003.

[107] A. Stancu *et al.*, "Investigation of the switching characteristics in ferroelectrics by first-order reversal curve diagrams," *Phys. B Condens. Matter*, vol. 372, no. 1–2, pp. 226–229, 2006.

[108] T. Schenk *et al.*, "Complex internal bias fields in ferroelectric hafnium oxide," *ACS Appl. Mater. Interfaces*, vol. 7, no. 36, pp. 20224–20233, Sep. 2015.

[109] L. Mitoseriu *et al.*, "First order reversal curves diagrams for describing ferroelectric switching characteristics," *Process. Appl. Ceram.*, vol. 3, pp. 2–7, 2009.

[110] J. E. Brewer and M. Gill, "Nonvolatile Memory Technologies with Emphasis on Flash A comprehensive guide to understanding and using NVM devices", *Hoboken, New Jersey: John Wiley & Sons, 2008*. 2008.

[111] J. Rodriguez *et al.*, "Reliability of Ferroelectric Random Access Memory embedded within 130 nm CMOS," *IEEE Int. Reliab. Phys. Symp. Proc.*, pp. 750–758, 2010.

[112] S.-J. Y. Kil *et al.*, "Development of New $TiN/ZrO_2/Al_2O_3/ZrO_2/TiN$ Capacitors Extendable to 45 nm Generation DRAMs Replacing HfO_2 Based Dielectrics," in *2006 Symposium on VLSI Technology, 2006. Digest of Technical Papers*, 2006, pp. 38–39.

[113] D. Zhou et al., "Reliability of Al_2O_3-doped ZrO_2 high-k dielectrics in three-dimensional stacked metal-insulator-metal capacitors," *J. Appl. Phys.*, vol. 108, p.124104, 2010.

[114] J. J. Park YK et al., "Fully Integrated 56 nm DRAM Technology for 1 Gb DRAM," in *2007 IEEE Symposium on VLSI Technology*, pp. 190–191, 2007.

[115] T. V. Perevalov et al., "Electronic structure of ZrO_2 and HfO_2." Defects in High-k Gate Dielectric Stacks," pp. 423–434, 2006.

[116] T. V. Perevalov et al., "Electronic structure of oxygen vacancies in hafnium oxide," *Microelectron. Eng.*, vol. 109, no. September, pp. 21–23, 2013.

[117] W. Weinreich et al., "Detailed leakage current analysis of metal–insulator–metal capacitors with ZrO_2, $ZrO_2/SiO_2/ZrO_2$, and $ZrO_2/Al_2O_3/ZrO_2$ as dielectric and TiN electrodes," *J. Vac. Sci. Technol. B Microelectron. Nanom. Struct.*, vol. 31, no. 2013, p. 01A109, 2013.

[118] G. C. Jegert et al., "Modeling of Leakage Currents in High-k Dielectrics", PhD disertation 2011.

[119] D. Martin et al., "Mesoscopic analysis of leakage current suppression in $ZrO_2/Al_2O_3/ZrO_2$ nano - laminates," *J. Appl. Phys.*, vol. 113, no. 19, p. 194103, 2013.

[120] G. Jegert et al., "Monte Carlo Simulation of Leakage Currents in $TiN/ZrO_2/TiN$ Capacitors," *IEEE Trans. Electron Devices*, vol. 58, no. 2, pp. 327–334, 2011.

[121] B. Kaczer et al., "Considerations for further scaling of metal – insulator – metal DRAM capacitors," *J. Vac. Sci. Technol.*, vol. B31, no. 1, p. 01A105, 2013.

[122] U. Schroeder et al., "Dielectric and metal electrode optimization for DRAM capacitor applications," 2015.

[123] S. Knebel et al., "Conduction Mechanisms and Breakdown Dielectrics for Three-Dimensional Stacked Metal – Insulator – Metal Capacitors," IEEE Transactions on Device and Materials Reliability vol. 14, no. 1, pp. 154–160, 2014.

[124] O. J. Kadoshima et al., "Effective - Work - Function Control by Varying the TiN Thickness in Poly-Si/TiN Gate Electrodes for Scaled High - k CMOSFETs," *IEEE Electron Device Lett.*, vol. 30, no. 5, pp. 466–468, 2009.

[125] S. Zafar et al., "A method for measuring barrier heights , metal work functions and fixed charge densities in metal/SiO_2/Si capacitors," vol. 4858, pp. 30–33, 2002.

[126] H. Wen et al., "Comparison of Effective Work Function Extraction Methods Using Capacitance and Current Measurement Techniques," vol. 27, no. 7, pp. 598–601, 2006.

[127] V. V. Afanas'ev, *Internal Photoemission Spectroscopy*. 2008.

[128] V.V. Afanas'ev, "Internal photoemission at interfaces of high-κ insulators with semiconductors and metals," *J. Appl. Phys.*, vol. 192, no. 8, p. 81301, 2007.

[129] S. Cimino et al., "A study of the leakage current in $TiN/HfO_2/TiN$ capacitors," *Microelectron. Eng.*, vol. 95, pp. 71–73, 2012.

[130] F. Jiménez-Molinos et al., "Direct and trap-assisted elastic tunneling through ultrathin gate oxides," *J. Appl. Phys.*, vol. 91, no. 8, pp. 5116–5124, 2002.

[131] A. Palma et al., "Quantum two-dimensional calculation of time constants of random telegraph signals in metal-oxide-semiconductor structures," *Physical.Review.B.(Condensed.Matter)*, vol. 56,

no. 15, pp. 9565–9574, 1997.

[132] D. R. Islamov et al., "Charge transport mechanism in thin films of amorphous and ferroelectric Hf$_{0.5}$Zr$_{0.5}$O$_2$," *JETP Lett.*, vol. 102, no. 8, pp. 544–547, 2015.

[133] D. R. Islamov et al., "Charge Carrier Transport Mechanism in High-k Dielectrics and Their Based Resistive Memory Cells," vol. 50, no. 3, pp. 310–314, 2014.

[134] S. Knebel et al., "Ultra-thin ZrO$_2$/SrO/ZrO$_2$ insulating stacks for future dynamic random access memory capacitor applications," *J. Appl. Phys.*, vol. 117, no. 22, Jun. 2015.

[135] R. Notes, "MDLab," p. 42122.

[136] O. Pirrotta et al., "Leakage current through the poly-crystalline HfO2: Trap densities at grains and grain boundaries," *J. Appl. Phys.*, vol. 114, no. 13, pp. 1–6, 2013.

[137] S. Mueller et al., "Ten-Nanometer Ferroelectric Si : HfO$_2$ Films for Next-Generation FRAM Capacitors," *1300 IEEE ELECTRON DEVICE Lett.*, vol. 33, no. 9, pp. 1300–1302, 2012.

[138] H. Mulaosmanovic et al., "Switching kinetics in nanoscale hafnium oxide based ferroelectric field effect transistors," *ACS Appl. Mater. Interfaces*, p. acsami.6b13866, 2017.

[139] X. Du et al., "Frequency spectra of fatigue of PZT and other ferroelectric thin films," vol. 493, pp. 311–316, 1998.

[140] Y. Kim et al., "Non-Kolmogorov-Avrami-Ishibashi switching dynamics in nanoscale ferroelectric capacitors," *Nano Lett.*, vol. 10, no. 4, pp. 1266–1270, 2010.

[141] Y.-H. Shin et al., "Nucleation and growth mechanism of ferroelectric domain-wall motion.," *Nature*, vol. 449, no. 7164, pp. 881–884, 2007.

[142] aixACCT Characterization tool user manual 2013.

[143] F. P. G. Fengler et al., "Comparison of hafnia and PZT based ferroelectrics for future non-volatile FRAM applications," pp. 369–372, 2016.

[144] N. Menou et al., "Polarization fatigue in PbZr$_{0.45}$Ti$_{0.55}$O$_3$ -based capacitors studied from high resolution synchrotron x-ray diffraction," *J. Appl. Phys.*, vol. 97, no. 6, pp. 1–7, 2005.

[145] D. Zhou et al., "Wake-up effects in Si-doped hafnium oxide ferroelectric thin films," *Appl. Phys. Lett.*, vol. 103, no. 19, 2013.

[146] X. J. Lou et al., "Polarization fatigue in ferroelectric thin films and related materials," *J. Appl. Phys.*, vol. 105, no. 2, 2009.

[147] V. Y. Shur et al., "Analysis of the Switching Data in Inhomogeneous Ferroelectrics," *Ferroelectrics*, vol. 349, no. 1, pp. 163–170, 2007.

[148] M. I. Morozov et al., "Hardening-softening transition in Fe-doped Pb(Zr,Ti)O$_3$ ceramics and evolution of the third harmonic of the polarization response," *J. Appl. Phys.*, vol. 104, no. 3, 2008.

[149] A. K. Tagantsev et al., "Interface-induced phenomena in polarization response of ferroelectric thin films," *J. Appl. Phys.*, vol. 100, no. 5, 2006.

[150] A. K. Tagantsev et al., "Polarization fatigue in ferroelectric films: Basic experimental findings, phenomenological scenarios, and microscopic features," *J. Appl. Phys.*, vol. 90, no. 3, pp. 1387–1402, 2001.

[151] W. L. Warren *et al.*, "Polarization suppression in Pb(Zr,Ti)O$_3$ thin films," vol. 6695, May, pp. 6695–6702, 1995.

[152] E. L. Colla *et al.*, "Direct observation of region by region suppression of the switchable polarization (fatigue) in Pb(Zr,Ti)O$_3$ thin film capacitors with Pt electrodes," *Appl. Phys. Lett.*, vol. 72, no. 21, pp. 2763–2765, 1998.

[153] A. K. Tagantsev *et al.*, "Identification of passive layer in ferroelectric thin films from their switching parameters," *J. Appl. Phys.*, vol. 78, no. 4, pp. 2623–2630, 1995.

[154] M. Pešic, *et al.*, "Impact of charge trapping on the ferroelectric switching behavior of doped HfO$_2$," *Phys. Status Solidi Appl. Mater. Sci.*, vol. 213, no. 2, pp. 270–273, Feb. 2016.

[155] T. Schenk *et al.*, "About the deformation of ferroelectric hystereses," *Appl. Phys. Rev.*, vol. 1, no. 4, p. 41103, 2014.

[156] A. L. Shluger *et al.*, "Models of Oxygen Vacancy Defects Involved in Degradation of Gate Dielectrics," pp. 1–9, 2013.

[157] S. R. Bradley *et al.*, "Electron-injection-assisted generation of oxygen vacancies in monoclinic HfO$_2$," *Phys. Rev. Appl.*, vol. 4, no. 6, pp. 1–7, 2015.

[158] N. Capron *et al.*, "Migration of oxygen vacancy in HfO$_2$ and across the HfO$_2$/SiO$_2$ interface: A first-principles investigation," *Appl. Phys. Lett.*, vol. 91, no. 19, p. 192905, 2007.

[159] K. Carl *et al.*, "Electrical after-effects in Pb(Ti, Zr)O$_3$ ceramics," *Ferroelectrics*, vol. 17, no. 1, pp. 473–486, 1977.

[160] M. Hoffmann *et al.*, "Low Temperature Compatible Hafnium Oxide Based Ferroelectrics," *Ferroelectrics*, vol. 480, no. 1, pp. 16–23, 2015.

[161] X. Sang *et al.*, "Revolving scanning transmission electron microscopy: Correcting sample drift distortion without prior knowledge," *Ultramicroscopy*, vol. 138, pp. 28–35, 2014.

[162] Y. Watanabe *et al.*, "Achievement of Higher-k and High-Φ in Phase Controlled HfO$_2$ Film using Post Gate-Electrode-Deposition Annealing," *ECS Trans.*, vol. 11, no. 4, pp. 35–45, 2007.

[163] D. Damjanovic *et al.*, "Ferroelectric, dielectric and piezoelectric properties of ferroelectric thin films and ceramics," *Reports Prog. Phys.*, vol. 61, no. 9, pp. 1267–1324, 1998.

[164] U. Schroeder *et al.*, "Impact of field cycling on HfO$_2$ based non-volatile memory devices," pp. 364–368, 2016.

[165] D. Starodub *et al.*, "Crystalline Oxides and Other High-K Materials on Silicon," no. 1, pp. 349–360.

[166] T. P. Ma *et al.*, "Why is nonvolatile ferroelectric memory field-effect transistor still elusive?," *IEEE Electron Device Lett.*, vol. 23, no. 7, pp. 386–388, 2002.

[167] E. Grimley *et al.*, "Structural Changes Underlying Field-Cycling Phenomena in Ferroelectric HfO$_2$ Thin Films," *Adv. Electron. Mater.*, 2016.

[168] P. McIntyre *et al.*, "Bulk and Interfacial Oxygen Defects in HfO$_2$ Gate Dielectric Stacks: A Critical Assessment," *ECS Trans.*, vol. 11, no. 4, pp. 235–249, 2007.

[169] H. Mulaosmanovic *et al.*, "Evidence of single domain switching in hafnium oxide based FeFETs: Enabler for multi-level FeFET memory cells," *Tech. Dig. - Int. Electron Devices Meet. IEDM*, vol. 2016-Febru, p. 26.8.1-26.8.3, 2016.

[170] C. Li et al., "Dynamic observation of oxygen vacancies in hafnia layer by in situ transmission electron microscopy," *Nano Res.*, vol. 8, no. 11, pp. 3571–3579, 2015.

[171] S. Starschich et al., "Evidence for oxygen vacancies movement during wake-up in ferroelectric hafnium oxide," *Appl. Phys. Lett.*, vol. 108, no. 3, 2016.

[172] L. Larcher et al., "Multi-scale modeling of HfO_x - ReRAM operation and variability RRAMs : state of the art." EMRS 2015.

[173] X. A. Tran et al., "Self-selection unipolar HfO_x-Based RRAM," *IEEE Trans. Electron Devices*, vol. 60, no. 1, pp. 391–395, 2013.

[174] M. Hoffmann et al., "Stabilizing the ferroelectric phase in doped hafnium oxide Stabilizing the ferroelectric phase in doped hafnium oxide," *J. Appl. Phys.*, vol. 118, no. 7, p. 72006, 2015.

[175] M. Pešić et al., "Conduction barrier offset engineering for DRAM capacitor scaling," *Solid. State. Electron.*, vol. 115, pp. 133–139, 2016.

[176] J. McPherson et al., "Thermochemical description of dielectric breakdown in high dielectric constant materials," *Appl. Phys. Lett.*, vol. 82, no. 13, pp. 2121–2123, 2003.

[177] M. Masuduzzaman et al., "Hot atom damage (HAD) limited TDDB lifetime of ferroelectric memories," *Tech. Dig. - Int. Electron Devices Meet. IEDM*, no. Fig 9, pp. 566–569, 2013.

[178] M. Masuduzzaman, et al., "Observation and control of hot atom damage in ferroelectric devices," *IEEE Trans. Electron Devices*, vol. 61, no. 10, pp. 3490–3498, 2014.

[179] U. Schroeder et al., "Impact of field cycling on HfO 2 based non-volatile memory devices." ESSDERC 2016.

[180] M. Grossmann et al., "The interface screening model as origin of imprint in $PbZr_xTi_{1-x}O_3$ thin films. I. Dopant, illumination, and bias dependence," *J. Appl. Phys.*, vol. 92, no. 5, pp. 2680–2687, 2002.

[181] I. Kanno et al., "Piezoelectric properties of *c*-axis oriented $Pb(Zr,Ti)O_3$ thin films," *Appl. Phys. Lett.*, vol. 70, no. 11, pp. 1378–1380, 1997.

[182] L. Hong et al., "White Paper Comparison of Embedded Non-Volatile Memory Technologies and Their Applications," no. May, pp. 1–8, 2009.

[183] U. Schroeder et al., *Materials for DRAM Memory Cell Applications. Thin Films on Silicon: Electronic and Photonic Applications.* 2016.

[184] J. Müller et al., "Ferroelectricity in simple binary ZrO_2 and HfO_2," *Nano Lett.*, vol. 12, no. 8, pp. 4318–4323, 2012.

[185] H. J. Kim et al., "Grain size engineering for ferroelectric $Hf_{0.5}Zr_{0.5}O_2$ films by an insertion of Al_2O_3 interlayer," *Appl. Phys. Lett.*, vol. 105, no. 19, pp. 0–6, 2014.

[186] "http://gwyddion.net/documentation/user-guide-en/." .

[187] G. Shirane et al., "Dielectric properties of lead zirconate," *Phys. Rev.*, vol. 84, no. 3, pp. 476–481, 1951.

[188] E. C. Park et al., "Effect of ion damage on the crystallization of PZT thin films," *Integr. Ferroelectr.*, vol. 31, no. 1–4, pp. 173–181, 2000.

[189] K. Kita et al., "Origin of electric dipoles formed at high-k/SiO_2 interface," *Appl. Phys. Lett.*, vol. 94, no. 13, p. 132902, 2009.

[190] H. K. Kim *et al.*, "Controlling work function and damaging effects of sputtered RuO_2 gate electrodes by changing oxygen gas ratio during sputtering," *ACS Appl. Mater. Interfaces*, vol. 5, no. 4, pp. 1327–1332, 2013.

[191] J. Hämäläinen *et al.*, "Atomic layer deposition of noble metals and their oxides," *Chem. Mater.*, vol. 26, no. 1, pp. 786–801, 2014.

[192] P. Pavan *et al.*, "Flash memory cells-an overview," *Proc. IEEE*, vol. 85, no. 8, pp. 1248–1271, 1997.

[193] F. P. G. Fengler *et al.*, "Domain Pinning: Comparison of Hafnia and PZT Based Ferroelectrics," *Adv. Electron. Mater.*, 2017.

BIBLIOGRAPHY

Resume

Name and surname:	MILAN D. PEŠIĆ
Date of birth:	14.04.1988
Country of birth:	SFR Yugoslavia

EDUCATION:

Since 01/2014	**Technical University of Dresden**, Faculty of Electrical and Computer Engineering (**PhD student**) Thesis topic: ELECTRICAL CHARACTERIZATION AND MODELING OF EMERGING POLARIZATION BASED NON-VOLATILE MEMORIES
02/2012-11/2013	**Technical University of Dresden**, Faculty of Electrical and Computer Engineering (**MSc Student**) Thesis topic: ELECTRICAL CHARACTERIZATION AND SIMULATION OF TRAPPING BEHAVIOR IN FERROELECTRIC FIELD EFFECT TRANSISTORS
10/2011-01/2012	**Technical University of Chemnitz**, Faculty of Electrical Engineering and Computer Science (**MSc Student**)
10/2007-07/2011	**University of Belgrade**, Faculty of Electrical Engineering and Computer Science (**BSc**) Thesis topic: SIMULATION OF THE ELECTRONIC STRUCTURE OF THE CORE SHELL NANOWIRES

WORK EXPERIENCE:

Since 11/2013	**Scientist** at **NaMLab gGmbH Dresden, Germany.** Topic: ELECTRICAL CHARACTERIZATION AND RELIABILITY OF NV MEMORIES AND DRAM; SIMULATION AND MODELING OF CHARGE TRANSPORT AND DEVICE PHYSICS; DEFECT CHARACTERIZATION
08/2015-09/2015	**Visiting researcher** at **MDLab** and **University of Modena and Reggio Emilia, Reggio Emilia, Italy** Topic: SIMULATION AND MODELING OF CHARGE/IONIC TRANSPORT AND TRAPPING WITHIN THE FeRAM NON-VOLATILE MEMORIES

List of Scientific Publications (from newest to oldest)

1) Fengler,F.P.G, **Pešić, M.**, Starschich, S., Schneller, T., Künneth, C., Böttger, U., Mulaosmanovic, H., Schenk, T., Park, MH., Nigon, R., Muralt, P., Mikolajick, T., and Schroeder, U., "Domain Pinning: Comparison of Hafnia and PZT Based Ferroelectrics" Accepted at Advanced Electronic Materials 2017.

2) **Pešić, M.**, Knebel, S., Hoffmann, M., Richter, C., Mikolajick, T. and Schroeder, U., How to Make DRAM non-volatile? Anti-ferroelectrics: A New Paradigm for Universal Memories Accepted at IEEE Transactions on Electron Devices. IEEE-IEDM 2016 will be published in 2017 IEEE International Electron Devices Meeting (IEDM) (pp. xx-yy). 2017.

3) Hoffmann, M., **Pešić, M.**, Khan, A. I., Salahuddin, S., Slesazeck, S. Schroeder, U., Mikolajick, T. "Direct Observation of Negative Capacitance in Polycrystalline Ferroelectric HfO_2". Advanced Functional Materials 26(47), 8643-8649, 2016.

4) **Pešić, M.**, Hoffmann, M., Richter, C., Mikolajick, T. and Schroeder, U. "Nonvolatile Random Access Memory and Energy Storage Based on Antiferroelectric Like Hysteresis in ZrO_2". Advanced Functional Materials 26(41), pp.7486-7494, 2016.

5) Schroeder, U., **Pešić, M.**, Schenk, T., Mulaosmanovic, H., Slesazeck, S., Ocker, J., Richter, C., Yurcuk, E., Khullar, K., Mueller, J., Polakowski, P., "Impact of field cycling on HfO_2 based non-volatile memory devices". 46th European Solid-State Device Research Conference (ESSDERC). pp.364-368, 2016.

6) Fengler, F. P.G, **Pešić, M.**, Starschich, S., Schneller, T., Böttger, T., Schenk, T. Park, M-H. Mikolajick, T., and Schroeder, U. "Comparison of hafnia and PZT based ferroelectrics for future non-volatile FRAM applications". 46th European Solid-State Device Research Conference (ESSDERC), pp.369- 372, 2016.

7) Yurchuk, E., Müller, J., Müller, S., Paul, J., **Pešić, M.**, van Bentum, R., Schroeder, U. and Mikolajick, T. "Charge-Trapping Phenomena in HfO_2-Based FeFET-Type Nonvolatile Memories". IEEE Transactions on Electron Devices, 63(9), pp.3501-3507, 2016.

8) Grimley, E. D., Schenk, T., Sang, X., **Pešić, M**, Schroeder, U., Mikolajick, T., LeBeau, J.M. "Structural Changes Underlying Field-Cycling Phenomena in Ferroelectric HfO_2 Thin Films". Advanced Electronic Materials, 2(9), 2016.

9) **Pešić, M.**, Fengler, F.P.G., Larcher, L., Padovani, A., Schenk, T., Grimley, E.D., Sang, X., LeBeau, J.M., Slesazeck, S., Schroeder, U. and Mikolajick, T. "Physical Mechanisms behind the Field- Cycling Behavior of HfO_2- Based Ferroelectric Capacitors". Advanced Functional Materials 26(25), 4601-4612. 2016.

10) **Pešić, M.**, Fengler, F.P., Slesazeck, S., Schroeder, U., Mikolajick, T., Larcher, L. and Padovani, A. "Root cause of degradation in novel HfO2-based ferroelectric memories". In Reliability Physics Symposium (IRPS), IEEE International (pp. MY-3). IEEE, 2016.

11) **Pešić, M.**, Knebel, S., Geyer, M., Schmelzer, S., Böttger, U., Kolomiiets, N., Afanas' ev, V.V., Cho, K., Jung, C., Chang, J. and Lim, H. "Low leakage ZrO2 based capacitors for sub 20 nm dynamic random access memory technology nodes". Journal of Applied Physics, 119(6), p.064101, 2016.

12) **Pešić, M.**, Knebel, S., Cho, K., Jung, C., Chang, J., Lim, H., Kolomiiets, N., Afanas'ev, V.V., Mikolajick, T. and Schroeder, U. "Conduction barrier offset engineering for DRAM capacitor scaling". Solid-State Electronics, 115, pp.133-139, 2016.

13) Mulaosmanovic, H., Slesazeck, S., Ocker, J., **Pešić, M.**, Muller, S., Flachowsky, S., Polakowski, P., Paul, J., Jansen, S., Kolodinski, S. and Richter, C... "Evidence of single domain switching in hafnium oxide based FeFETs: Enabler for multi-level FeFET memory cells". In IEEE International Electron Devices Meeting (IEDM) (pp. 26-8). IEEE, December 2015.

14) **Pešić, M**., Slesazeck, S., Schenk, T., Schroeder, U. and Mikolajick, T. "Impact of charge trapping on the ferroelectric switching behavior of doped HfO_2". physica status solidi (a), 2015.

15) Schenk, T., Hoffmann, M., Ocker, J., **Pešić, M**., Mikolajick, T. and Schroeder, U. "Complex internal bias fields in ferroelectric hafnium oxide". ACS applied materials & interfaces, 7(36), pp.20224-20233, 2015.

16) Knebel, S., **Pešić, M**., Cho, K., Chang, J., Lim, H., Kolomiiets, N., Afanas' ev, V.V., Muehle, U., Schroeder, U. and Mikolajick, T. "Ultra-thin $ZrO_2/SrO/ZrO_2$ insulating stacks for future dynamic random access memory capacitor applications". Journal of Applied Physics, 117(22), p.224102, 2015.

17) Cho, K., **Pešić, M**., Knebel, S., Jung, C., Chang, J., Lim, H., Kolomiiets, N., Afanas' ev, V.V., Schroeder, U. and Mikolajick, T., "Schottky barrier height engineering for next generation DRAM capacitors". In IEEE Joint International EUROSOI Workshop and International Conference on Ultimate Integration on Silicon (EUROSOI-ULIS), (pp. 129-132). January, 2015.

18) Schenk, T., Schroeder, U., **Pešić, M**., Popovici, M., Pershin, Y.V. and Mikolajick, T. "Electric field cycling behavior of ferroelectric hafnium oxide". ACS applied materials & interfaces, 6(22), pp.19744-19751, 2014.

Conference Talks

1) IEEE International Electron Devices Meeting **IEEE-IEDM 2016**, At San Francisco, CA, USA. How to Make DRAM non-volatile? Anti-ferroelectrics: A New Paradigm for Universal Memories.

2) International Reliability Physics Symposium **IEEE-IRPS 2016**, At Pasadena, CA, USA. Root cause of degradation in novel HfO_2-based Ferroelectric Memories.

3) European Material Research Society **EMRS 2015**-Lille, France. Modeling of Charge Trapping influence on the Ferroelectric Switching Behavior of Doped HfO_2.

4) **DPG Spring Meeting 2014**. Modeling of Influence of Charge Trapping on Memory Characteristics of $Si:HfO_2$-Based Ferroelectric Field Effect Transistors.

5) 4rd International Workshop on Simulation and Modeling of Memory Devices (**IWSMM**), 2013 Influence of Charge Trapping on Memory Characteristics of $Si:HfO_2$-Based Ferroelectric Field Effect Transistors. (*as a master student*)

Conference Contributions

1) 46th European Solid-State Device Research Conference (**ESSDERC 2017**), 2017, At Leuven, BE. M. Pešić, M. Hoffmann, C. Richter, S. Slesazeck, T. Kämpfe, L. M. Eng, T. Mikolajick, U. Schroeder. Anti-ferroelectric ZrO_2: An Enabler for Low Power Non-volatile 1T-1C and 1T Random Access Memories.

2) Joint International EUROSOI Workshop and International Conference on Ultimate Integration on Silicon (**EUROSOI-ULIS 2017**), 2017, At Athens, GRE. M. Hoffmann, **M. Pešić**, S. Slesazeck, U. Schroeder, T. Mikolajick. Modeling and Design Considerations for Negative Capacitance Field-Effect Transistors.

3) 9th International Memory Workshop (**IMW 2017 Tutorial**), 2017, At Monterey, CA, US. U. Schroeder, **M. Pešić**, S. Knebel, M. Hoffmann, C. Richter, E. Grimley, J. LeBeau, T. Mikolajick. HfO_2 based FeRAM and NVDRAM.

4) 9th International Memory Workshop (**IMW 2017 Tutorial**), 2017, At Monterey, CA, US. U. Schroeder, T. Schenk, M. H. Park, C. Richter, M. Hoffmann, **M. Pešić**, J. Müller, T. Shimizu, H. Funakubo, D. Pohl, R. Materlik, A. Kersch, E. Grimley, J. LeBeau, J. Jones, N. Wisinger, S. Kalinin, and T. Mikolajick. Ferroelectric HfO_2: Basics, material properties and optimization.

5) Joint IEEE International Symposium on the Applications of Ferroelectrics, European Conference on Application of Polar Dielectrics, and Piezoelectric Force Microscopy Workshop (**ISAF/ECAPD/PFM 2017**) Atlanta, GE, US. **M. Pešić**, M. Hoffmann, C. Richter, S. Slesazeck, T. Mikolajick,U. Schroeder. Anti-ferroelectric HfO_2 or ZrO_2:A key material for novel anti-ferroelectric non-volatile memories.

6) Towards oxide based electronics (**TO-BE 2017**), 2017 Luxembourg, LUX. T. Schenk, M. H. Park, **M. Pešić**, M. Hoffmann, C. Richter, S. Mueller, H. Mulaosmanovic, F. P. G. Fengler, S. Slesazeck, T. Mikolajick, U. Schroeder. 10 Years Fluorite-type Ferroelectrics – A Survey.

7) The 48th IEEE Semiconductor Interface Specialists Conference (**SISC 2016**), 2016, At San Diego, CA, USA. J. Müller, M. Trentzsch, S. Flachowsky, H. Mulaosmanovic, S. Müller, P. Polakowski, R. Richter, J. Paul, B. Reimer, D. Utess, S. Jansen, J. Schreiter, J. Ocker, M. Noack, S. Riedel, **M. Pesic**, S. Kolodinski, U. Schröder, S. Slesazeck, T. Mikolajick, S. Beyer. Material Innovation Ferroelectric Hafnium Oxide: Towards Cheaper Memories, Steeper Slopes and New Value Adders for HKMG.

8) IEEE International Electron Devices Meeting (**IEEE-IEDM 2016**), At San Francisco, CA, USA. **M Pesic**, S Knebel, M Hoffmann, C Richter, T Mikolajick, U Schroeder. How to Make DRAM non-volatile? Anti-ferroelectrics: A New Paradigm for Universal Memories.

9) 46th European Solid-State Device Research Conference (**ESSDERC 2016**), 2015, At Lausanne. F. P.G. Fengler, **M. Pešić**, S. Starschich, T. Schneller, U. Böttger, T. Schenk, M. H. Park, T. Mikolajick, U. Schroeder. Comparison of hafnia and PZT based ferroelectrics for future non-volatile FRAM applications.

10) 46th European Solid-State Device Research Conference (**ESSDERC 2016**), 2015, At Lausanne. U. Schroeder, **M Pešic**, T. Schenk, H Mulaosmanovic, S. Slesazeck, J Ocker, C Richter, E Yurchuk, K Khullar, J Müller, P Polakowski, ED Grimley, JM LeBeau, S Flachowsky, S Jansen, S Kolodinski, R van Bentum, A Kersch, C Künneth, T. Mikolajick. Impact of field cycling on HfO_2 based non-volatile memory devices.

11) Joint IEEE International Symposium on the Applications of Ferroelectrics, European Conference on Application of Polar Dielectrics, and Piezoelectric Force Microscopy Workshop (**ISAF/ECAPD/PFM 2016**) Darmstadt, GER. F. Fengler, **M. Pešić**, S. Starschich, T. Schneller, U. Böttger, L. Larcher, A. Padovani, T. Schenk, M.H. Park, E.D. Grimley, J.M. LeBeau,T. Mikolajick, and U. Schroeder. Wake-up behavior comparison between PZT and HfO_2 based ferroelectrics.

12) Joint IEEE International Symposium on the Applications of Ferroelectrics, European Conference on Application of Polar Dielectrics, and Piezoelectric Force Microscopy Workshop (**ISAF/ECAPD/PFM 2016**) Darmstadt, GER. T. Schenk, M. Hoffmann, J. Ocker, **M. Pešić**, E. D. Grimley, X. Sang, J. M. LeBeau, T. Mikolajick, and U. Schroeder. Internal Bias Fields in Ferroelectric HfO_2 Thin Films and their Structural Origins.

13) **CIMTECH** conference 2016, Perugia, IT. U. Schroeder, T. Schenk, M. Hoffmann, C. Richter, **M. Pešić**, F. Fengler, M.H. Park, S. Slesazeck, D. Pohl, C. Künneth, R. Materlik, A. Kersch, X. Sang, E. Grimley, J. LeBeau, N. Wisinger, S. Kalinin, and T. Mikolajick. Ferroelectric HfO_2 for Non-Volatile Memory Devices.

14) AVS-ALD conference 2016, Dublin, IR. U. Schroeder, T. Schenk, C. Richter, **M. Pešić**, F. Fengler, M.H. Park, S. Slesazeck, D. Pohl, C. Künneth, R. Materlik, A. Kersch,X. Sang, E. Grimley, J. LeBeau, N. Wisinger, S. Kalinin, and T. Mikolajick. Impact of ALD processing on non-volatile memory performance of ferroelectric HfO_2 based capacitors.

15) IEEE International Electron Devices Meeting **IEEE-IEDM 2015**, At Washington, D.C, USA. H Mulaosmanovic, S Slesazeck, J Ocker, **M Pesic**, S Muller, S Flachowsky, J Müller, P Polakowski, J Paul, S Jansen, S Kolodinski, C Richter, S Piontek, T Schenk, A Kersch, C Kunneth, R van Bentum, U Schroder, T Mikolajick Evidence of single domain switching in hafnium oxide based FeFETs: Enabler for multi-level FeFET memory cells.

16) Joint International EUROSOI Workshop and International Conference on Ultimate Integration on Silicon (**EUROSOI-ULIS 2015**), 2015, At Bologna, IT. K. Cho, **M. Pešić**, S. Knebel, C. Jung, J. Chang, H. Lim, N. Kolomiiets, V. V. Afanas' ev, U. Schroeder, T. Mikolajick. Schottky barrier height engineering for next generation DRAM capacitors.

17) European Material Research Society (**EMRS 2015**)-Lille, France. **M. Pešić**, S. Slesazeck, T. Schenk, T., Schroeder, U. and Mikolajick, T Modeling of Charge Trapping influence on the Ferroelectric Switching Behavior of Doped HfO_2.

18) Material Research Society Meeting (**MRS 2015**)-Boston, MA, US. T. Schenk, M. Hoffmann, C. Richter, **M. Pešić**, S. V. Kalinin, A. Kersch, T. Mikolajick, and U. Schroeder. Doped Hafnium Oxide for Ferroelectric Memories.

19) 4rd International Workshop on Simulation and Modeling of Memory Devices (**IWSMM**), 2013 Influence of Charge Trapping on Memory Characteristics of Si:HfO_2-Based Ferroelectric Field Effect Transistors. (*as a master student*)

Patent application

1) US 15/057,884

BIBLIOGRAPHY

Herstellung und Verlag:
BoD - Books on Demand, Norderstedt
ISBN 978-3-7448-6788-7

www.ingramcontent.com/pod-product-compliance
Lightning Source LLC
Chambersburg PA
CBHW070244230526
45470CB00002B/477